Burning Swamps

This groundbreaking environmental history recounts the story of Russia's fossil economy from its margins. Unpacking the forgotten history of how peat fueled manufacturing industries and power plants in late imperial and Soviet Russia, Katja Bruisch provides a corrective to more familiar historical narratives dominated by coal, oil, and gas. Attentive to the intertwined histories of matter and labor during a century of industrial peat extraction, she offers a fresh perspective on the modern Russian economy that moves beyond the socialism/capitalism binary. By identifying peat extraction in modern Russia as a crucial chapter in the degradation of the world's peatlands, Bruisch makes a compelling case for paying attention to seemingly marginal places, people, and resources as we tell the histories of the planetary emergency.

Katja Bruisch is Ussher Assistant Professor in Environmental History at Trinity College Dublin.

Studies in Environment and History

Other Books in the Series

Melanie Arndt *Chernobyl Children: A Transnational History of a Nuclear Disaster*
Sverker Sörlin and Eric Paglia *Stockholm and the Rise of Global Environmental Governance: The Human Environment*
Zozan Pehlivan *The Political Ecology of Violence: Peasants and Pastoralists in the Last Ottoman Century*
Matthew P. Johnson *Hydropower in Authoritarian Brazil: An Environmental History of Low-Carbon Energy, 1960s–1990s*
Ellen Arnold *Medieval Riverscapes: Environment and Memory in Northwest Europe, c. 300–1100*
Richard C. Hoffmann *The Catch: An Environmental History of Medieval European Fisheries*
Samuel Dolbee *Locusts of Power: Borders, Empire, and Environment in the Modern Middle East*
Andy Bruno *Tunguska: A Siberian Mystery and Its Environmental Legacy*
Lionel Frost *Cities in a Sunburnt Country: Water and the Making of Urban Australia*
Adam Sundberg *Natural Disaster at the Closing of the Dutch Golden Age: Floods, Worms, and Cattle Plague*
Germán Vergara *Fueling Mexico: Energy and Environment, 1850–1950*
Peder Anker *The Power of the Periphery: How Norway Became an Environmental Pioneer for the World*
David Moon *The American Steppes: The Unexpected Russian Roots of Great Plains Agriculture, 1870s–1930s*
James L. A. Webb, Jr. *The Guts of the Matter: A Global Environmental History of Human Waste and Infectious Intestinal Disease*
Maya K. Peterson *Pipe Dreams: Water and Empire in Central Asia's Aral Sea Basin*
Thomas M. Wickman *Snowshoe Country: An Environmental and Cultural History of Winter in the Early American Northeast*
Debjani Bhattacharyya *Empire and Ecology in the Bengal Delta: The Making of Calcutta*
Chris Courtney *The Nature of Disaster in China: The 1931 Yangzi River Flood*
Dagomar Degroot *The Frigid Golden Age: Climate Change, the Little Ice Age, and the Dutch Republic, 1560–1720*

(*Continued after the Index*)

Burning Swamps

Peat and the Forgotten Margins of Russia's Fossil Economy

KATJA BRUISCH
Trinity College Dublin

CAMBRIDGE
UNIVERSITY PRESS

CAMBRIDGE
UNIVERSITY PRESS

Shaftesbury Road, Cambridge CB2 8EA, United Kingdom

One Liberty Plaza, 20th Floor, New York, NY 10006, USA

477 Williamstown Road, Port Melbourne, VIC 3207, Australia

314–321, 3rd Floor, Plot 3, Splendor Forum, Jasola District Centre, New Delhi – 110025, India

103 Penang Road, #05–06/07, Visioncrest Commercial, Singapore 238467

Cambridge University Press is part of Cambridge University Press & Assessment, a department of the University of Cambridge.

We share the University's mission to contribute to society through the pursuit of education, learning and research at the highest international levels of excellence.

www.cambridge.org
Information on this title: www.cambridge.org/9781009603089

DOI: 10.1017/9781009603096

When citing this work, please include a reference to the DOI 10.1017/9781009603096

First published 2025

Cover image: Peat worker in the Ural region, 1920s. Photo: Russian State Film and Photo Archive (RGAKFD). Reproduced with permission.

A catalogue record for this publication is available from the British Library

A Cataloging-in-Publication data record for this book is available from the Library of Congress

ISBN 978-1-009-60308-9 Hardback

To Anna and Rubén

Contents

Figures

Acknowledgments

Many people and institutions helped me make this book possible. My time at the German Historical Institute in Moscow, once a place of academic exchange and now an "undesired organization" in Russia, marked an important stage in my personal and academic trajectory. I was fortunate to work there while Nikolaus Katzer and later Sandra Dahlke led the institute with much humor and a deep sense of responsibility. Maria Golovnya, who passed away far too young, would have ironically raised her eyebrow had I told her I would mention her unforgettable laughter in these acknowledgments. Larisa Kondrat'eva, Victoria Silwanowitsch and Matthias Uhl organized copies of library sources years after I left Moscow in 2016. Irina Lukka from the Slavonic Library in Helsinki did the same when the COVID-19 pandemic interrupted my long-anticipated stay in Finland in March 2020.

I knew little about peat or peatlands when I first started, and I was fortunate to get great advice and good company during my forays into their history and ecology. Between 2016 and 2019, I traveled to several former peat extraction sites and workers' settlements near Moscow, Saint Petersburg and Yaroslavl. I feel greatly indebted to the colleagues and friends without whom I would have never been able to undertake these memorable trips that allowed me to picture the places whose histories I studied: Igor' Kuznetsov, Ivan Mikirtumov, Aleksandr Nikulin, Aleksandr Vatlin, and Ruslan Tsvygun. I also thank everyone willing to show us directions or talk to us while we explored the forgotten margins of Russia's fossil economy. In the spring of 2015, when I was just conceiving this project, I visited the Greifswald Mire Centre, where I got to know some of the exciting work of Greifswald's peatland scientists.

Conversations with Hans Joosten at the wooden table of his astounding library cleared my doubts about whether the history of peat was a worthy subject. I was fortunate to return to Greifswald in April 2024 when this book neared completion. Hans Joosten and Franziska Tanneberger made connections for me and showed me the beauty of rewetted peatlands. During this visit, I was fortunate to meet Cosima Tegetmeyer, who made a wonderful map for this book.

My journey into the past of central Russia's peatlands would have been impossible without the material and intellectual support I received over the years. Funding from the Royal Irish Academy, the Aleksanteri Institute in Helsinki, the Grace Lawless Lee Fund and the AHSS Benefactions Fund at Trinity College Dublin allowed me to conduct archival and library research. In the summer of 2022, I spent several weeks at the GWZO in Leipzig, where I enjoyed some quiet writing time and inspiring conversations with colleagues curious about peat and other matters. Presentations, most of them virtual, at the Jordan Centre at New York University, the Polish Academy of Sciences, the Higher School of Economics in Saint Petersburg, Newcastle University, the University of Bern, the Free University Berlin, the Imre Kertész Kolleg Jena, the University of Freiburg, the Berlin–Brandenburg Kolloquium für Umweltgeschichte and the Energy History Working Group allowed me to probe my ideas with generous scholars of various backgrounds.

It is hard to express my gratitude to everyone who engaged with my writing once the manuscript gained contours. At different stages of the process, Nathan Kapoor, Michael Kopsidis, Daniela Russ, and Lily Toomey provided thoughtful and constructive feedback that helped me improve my work. If there is one positive legacy of the COVID-19 pandemic, it is the Zoom *kruzhok,* where I got so much inspiration and encouragement since early 2021. It means a lot to me that Michel Abesser, Boris Belge, Maike Lehmann, Alex Oberländer, Immo Rebitschek, and Alexa von Winning spent many hours reading and discussing large parts of this manuscript. There is no doubt that their feedback made this a better book. Special thanks to Alex for her virtual presence during uncounted writing days and her tireless, warm-hearted cheering over the years. Talking theory and life with Diogo de Carvalho Cabral and getting his views on my writing has been an incredible privilege and shaped the way I think. Jennifer Keating's astute comments on several chapters, reassurance, and friendship made me trust that there was meaning in my trying.

I was lucky to have dedicated people around me who supported me during the editing and publishing process. Rob Browne thoroughly proofread the first complete version of the book and proposed stylistic improvements. Jamie Rohu provided editorial assistance when the manuscript neared completion. Without Sarah Smyth's help, I would have struggled to retain the spirit of peat workers' folklore in English translation. I thank the two peer reviewers, one of whom turned out to be Andy Bruno, for their generous and constructive engagement with my work. At Cambridge University Press, it was thanks to the professionalism and guidance of Lucy Rhymer and Rosa Martin that I always felt this book was in the right hands. Thanks also to Liz Steel, my copy-editor, as well as Clare Dennison and Gabriel Sam Antony for overseeing the production of this book. Ben Shaw, who compiled the index, was an absolute delight to work with. In a section of chapter 6, I draw on my "How Wetlands entered the Transnational Spaces of Late Soviet Environmentalism" in *Thinking Russia's History Environmentally* (2023), edited by Catherine Evtuhov, Julia Lajus and David Moon. I thank Berghahn for the permission to reuse this material.

While all makes sense with hindsight, I could have never imagined that I would once live in Ireland. At Trinity College Dublin, I have enjoyed the conversations with and companionship of Carole Holohan, Isabella Jackson, Georgina Laragy, Molly Pucci, and Clare Tebbutt. Several sections of this book were written in the calming presence of Molly's purring cats. Thank you to Jane Suzanne Carroll for the Golden columbine. At the Trinity Centre for Environmental Humanities, I became aware of fields of research that I didn't even know existed. I was lucky to have Ruth Brennan and Cordula Scherer near me for some much-needed distraction. While working on the manuscript, I had the great pleasure to supervise PhD students Gabriel Coleman and Lily Toomey, whose vision and creativity keep inspiring me. Celeste Golden, Ronan Foley, Joe O'Sullivan, and Nell Reagan made my family feel welcome in Shankill. The Shanganagh Community Gardens have been a spring of joy throughout the seasons – it is this colorful community of growers, plants, and little creatures nestled between the railways and a wastewater treatment plant that gives me hope for life in our precarious times. To Rubén Flores Sandoval and Anna, who wished this were a children's book, I owe what makes my days rich and bright. I dedicate this book to them.

A Note on Usage

TRANSLATIONS AND TRANSLITERATIONS

Translations are my own, except for published translations of literary works. Sarah Smyth assisted me in my efforts to retain the spirit of the peat workers' folklore cited in Chapters 3 and 4 in English translation. I acknowledge her contribution in the notes. Transliterations follow the Library of Congress system. When using Russian terms and units of measure in the text, I leave out apostrophes that usually signify the Cyrillic soft sign (artel instead of *artel'*) and anglicize the plural (desiatinas instead of *desiatiny*) to enhance readability. For the same reason, the text uses simplified common English versions for names and toponyms, so it will be Trotsky instead of Trotskii, Ryazan instead of Riazan'. I follow the Library of Congress transliteration system in the notes and the bibliography.

GLOSSARY

artel'	voluntary association of people for joint work; a widespread form of organization among peasants in the Russian Empire who left their village for temporary outside work
desiatina	unit of area equivalent to 1.0925 hectares
GOĖLRO	State Commission for the Electrification of Russia
Gosplan	State Planning Committee
GUZiZ	Main Administration of Land Management and Agriculture

IUCN	International Union for Conservation of Nature
kolkhoz	collective farm
Narkomzem	People's Commissariat of Agriculture
NEP	New Economic Policy
pud	unit of weight equivalent to 16.3807 kg
RSFSR	Russian Soviet Federated Socialist Republic
sazhen'	unit of length equivalent to 2.1336 m
sovkhoz	state farm
Trudy VĖO	Transactions of the Free Economic Society
UNESCO	United Nations Educational, Scientific and Cultural Organization
verst	unit of distance equivalent to 1.0668 km
VSNKh	Supreme Council of the National Economy
zemstvo	organ of self-government established at provincial and district level in 1864 as part of Alexander II's reforms

ARCHIVES AND ARCHIVAL TERMS

ARAN	Archive of the Russian Academy of Sciences
GARF	State Archive of the Russian Federation
RGAĖ	Russian State Archive of the Economy
RGIA	Russian State Historical Archive
d. (delo)	file
f. (fond)	collection
l., ll. (listok, listy)	page, pages
ob. (oborot)	verso
op. (opis')	register

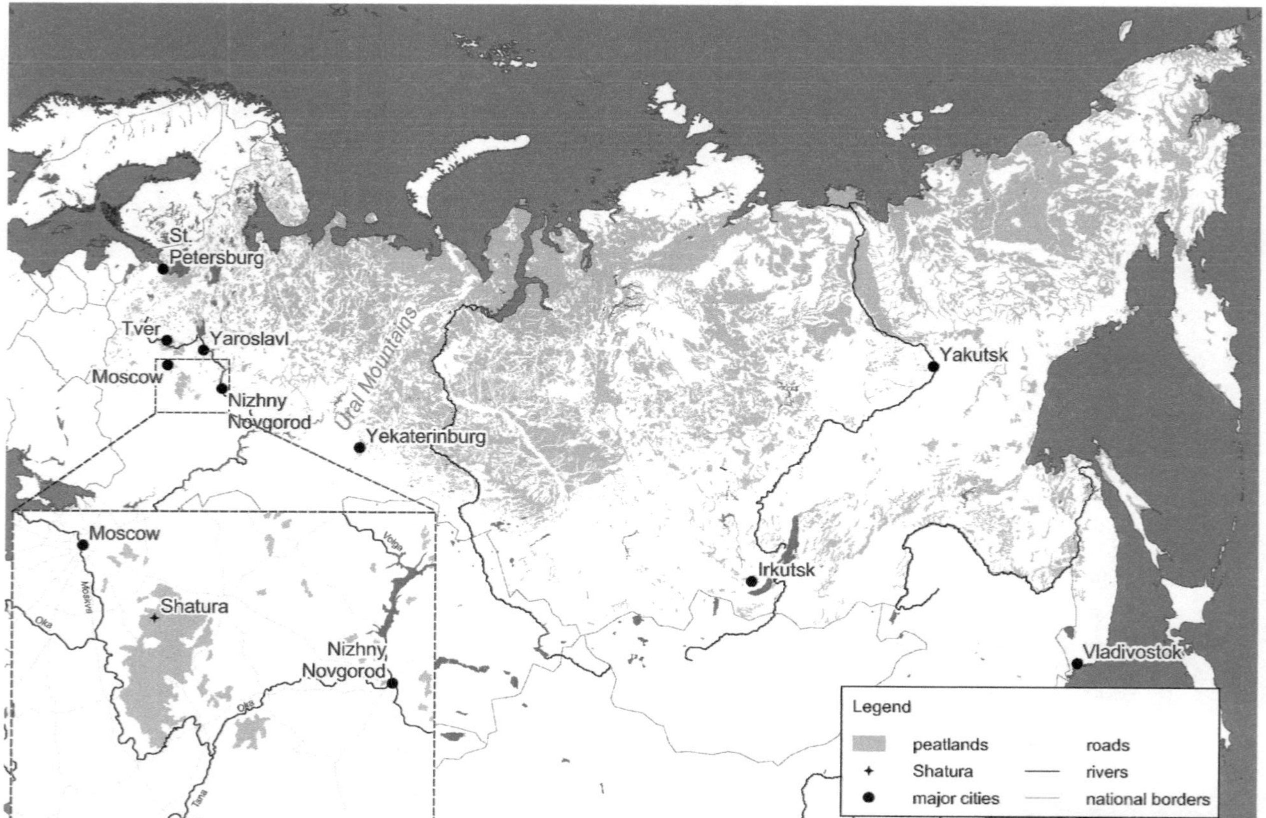

FIGURE I.1 Map of peatlands in the Russian Federation and in the region of Shatura. Based on data from the Global Peatland Database / Greifswald Mire Centre (2024). Peatland distribution in Russia after Vompersky et al. (2011). Map by Cosima Tegetmeyer. Made with Natural Earth.

Introduction

My first conscious encounter with Russia's peatlands was one of immediacy and distance at the same time. In the summer of 2010, prolonged heat and drought caused massive wildfires across central Russia. I had been looking forward to traveling to Moscow, but excitement gave way to unease once I arrived in early August. The atmosphere was eerie when I left the airport, as the wind blew the smoke from regions further east, where the fires burnt ferociously, into the city. I had read about the situation, but words and images left me unprepared for the experience that awaited me. Some days the smoke was so thick that it was difficult to make out the houses across the street. The smoke crept through the wooden window frames of the apartment where I was staying. Breathing was difficult. Sleeping even more so. I developed a cough and bought a mask to cover my nose and mouth. But my daily discomfort was trivial compared to the displacement and death the fires brought beyond the Russian capital. The village of Mokhovoe, some 150 kilometers to Moscow's southeast, was erased within minutes during that summer. The entire population lost their homes. Twelve locals perished. Mokhovoe was never rebuilt, while its former inhabitants were moved into hastily constructed standardized houses in a larger settlement nearby. Leaving a landscape of charred ruins and dead trees, they carried a trauma bound to stay with them.[1]

[1] Irina Reznik, "Derevnia sgorela ugolovno" (September 13, 2010), www.gazeta.ru/social/2010/09/13/3419046.shtml. Accessed July 15, 2024; Sergei Liutykh, "'Tela liudei ne goreli, a raspukhali ot temperatury': 10 let nazad Moskva zadykhalas' ot smoga, a vokrug sgorali tselye derevni. Mozhet li eto povtorit'sia?" (July 27, 2020), https://lenta.ru/articles/2020/07/27/fire2010/. Accessed July 15, 2024.

The uncanniness of that summer remained with me long after the inferno was over. At the time, experts had pointed to the underfunded fire services, which compounded the fire vulnerability facing Russia's forests as periods of drought and heat became more severe and frequent. Wildfires in Russia continuously worsened during the 1990s and even more so after the 2005 Forest Code relegated fire prevention and suppression to poorly resourced regional administrations.[2] While the intersection between climate change and the immediate political context is crucial to understanding the summer of 2010, the nature of the disaster also hints at deeper layers of the past. In the eastern and southeastern parts of the Moscow region, the fires spread through a landscape marked by extensive drainage and peat extraction in the twentieth century. Journalists within the country and abroad drew a link between the 2010 emergency and the dire state of central Russia's exploited peatlands. Titled "Past errors to blame for Russia's peat fires," a *New York Times* article suggested a direct connection between Lenin's electrification campaign, the drainage of peatlands, and the smoke that clouded Moscow's skies almost a century after the 1917 Revolution.[3] None of this was immediately apparent to me. Although Russia is the country with the largest known peatland area in the world, I had not encountered peatlands in my history books. Nor had I considered that peat could be a matter for historians. Past and present spoke to each other in ways I struggled to understand.[4]

It took me years to accept this irritation as an invitation. Once I did, I learned that peat had been used as an industrial fuel long before the

[2] Ivan Blokov, "Alle Jahre wieder: Waldbrände in Russland," *Osteuropa* 60, no. 9 (2010): 7–13.

[3] "Moskva v dymu: Torfianye pozhary podkhodiat k gorodu" (July 28, 2010), www.rbc.ru/society/28/07/2010/5703dc6d9a79470ab5023322. Accessed July 15, 2024; Alister Doyle, "Russia's Peat Fires Seen Burning For Months" (August 12, 2010), www.reuters.com/article/us-russia-heat-peat-idUSTRE67A3H120100812. Accessed July 15, 2024; Andrew E. Kramer, "Past Errors to Blame for Russia's Peat Fires," *New York Times* (August 12, 2010), www.nytimes.com/2010/08/13/world/europe/13russia.html. Accessed July 15, 2024.

[4] For an earlier reflection on my journey into environmental history, see Katja Bruisch, "Smouldering Past: How a Fire Event Turned Me into an Environmental Historian," NiCHE (August 31, 2022), https://niche-canada.org/2022/08/31/smouldering-past-how-a-fire-event-turned-me-into-an-environmental-historian/. Accessed March 18, 2025.

Bolshevik takeover and that the Soviet peat industry had had its own bureaucracy. Reading what archives and libraries had to offer, I began to look up places from the written record on Google Maps, where a geography of extractivism and settlement, transport and drainage infrastructure, fragmentation and connectivity emerged before my eyes. In 2016, I went on my first trip to the Shatura region. The town of Shatura had occupied an important place in the Soviet electrification myth since the opening of its regional power plant in 1925, a symbol of the utopian spirit and mobilizing power of the Soviet project. Until it was converted to run primarily on natural gas in the late 1980s, Shatura's power plant received its fuel from the surrounding peatlands. Mokhovoe, wiped from the map in 2010, had been located close to abandoned extraction sites from where peat used to be shipped to Shatura. Peat extraction had ceased decades before I visited, but its former importance resonated strongly. The power plant was still working, its chimneys towering over the leafy streets of the town. In front of Shatura's Museum of Local Lore, the monument of a woman carrying a basket full of peat evoked a strong connection between resource extraction, labor, and the town's identity (Figure I.2). In a workers' settlement a few kilometers outside Shatura, three women shared memories of a time when life in the area had revolved around peat. Their stories made me realize how closely peat and people's sense of home were tied together in this part of Russia. The places of the written record may not provide us with the answers we are looking for, but they can hold clues that are easy to miss in the two-dimensional reality created by archival documents and library holdings.[5]

Burning Swamps recounts the history of Russia's modern economy from its margins. A journey through the expansive peatlands of central Russia, it tells the history of a forgotten fossil energy carrier that once fueled industries and power plants, of workers who exploited their bodies for the energy needs of an expanding economy, of place-making through extraction, and of wetlands that have become highly vulnerable to fire – a danger that only increases as the world gets warmer. Reconstructing

[5] Kate Brown, *Dispatches from Dystopia: Histories of Places Not Yet Forgotten* (Chicago: The University of Chicago Press, 2015).

FIGURE I.2 Monument devoted to female peat workers in Shatura.
Source: Author's photograph (2016).

peat's role as a regional energy source and the socio-ecological impact of its extraction, this book claims that Russia's energy system was historically much more diverse than top-down narratives of the fossil fuel age suggest. At the same time, it identifies the extraction and use of peat in Russia as a crucial chapter in the global history of how the energy demands of industrial societies made peatlands and the people exploiting them subject to the imperative of fuel extraction. In places like Shatura and its surroundings, this history was full of ambiguity. As wetlands and workers became part of Russia's industrial metabolism, the regions shaped by the extraction of peat turned into sites of contestation and resistance as much as home and belonging. It is this tension – between the destruction and the creation of life, loss and possibility, inclusion and marginalization – that lies at the heart of this book.

Revisiting the historical geographies of energy and growth in Eurasia, *Burning Swamps* underscores the significance of lesser-known resource histories for understanding the planetary crises of the present day. Ranked among the top five greenhouse gas emitters globally, Russia is also among the world's major climate debtors, which means that its historical emissions have exceeded what would be considered a "fair" share in the "atmospheric commons" if humans were to live within planetary boundaries.[6] From the late nineteenth century, Russia's economic trajectory combined industrial expansion with an increasing reliance on fossil fuels. Imperial domination over territories rich in fossil energy resources, such as the Donets coal basin in Ukraine and the oil fields in the South Caucasus, was crucial for the empire's industrial take-off. Siberian fossil fuels became relevant in the second half of the twentieth century. But Russia's rising industrial metabolism and its increasingly excessive combustion of fossil energy sources also hinged upon rural regions close to political and urban metropoles, where the extraction and burning of peat turned wetlands into wastelands and released massive amounts of carbon dioxide into the atmosphere. Attending to these places means recognizing the multiple forms of resource extractivism in Russia's empires and the diverse historical pathways leading into the climate emergency.

The forgotten role of peat in Russia's economy is an example of how economic growth transformed the global countryside, as human labor, minerals, and land in rural hinterlands and imperial peripheries became resources for commodity production and urban development.[7] Histories

[6] In 2022, Russia was the world's fifth largest greenhouse gas emitter. M. Crippa et al., *GHG Emissions of All World Countries* (Luxembourg: Publications Office of the European Union, 2023), 16, doi: 10.2760/953322, JRC134504. For a quantitative perspective on national historical carbon debt and a discussion of nations' "fair share," see Jason Hickel, "Quantifying National Responsibility for Climate Breakdown: An Equality-based Attribution Approach for Carbon Dioxide Emissions in Excess of the Planetary Boundary," *The Lancet Planetary Health* 4, no. 9 (2020): e399–e404. A similar attempt, though going back only to the 1960s, was made by H. Damon Matthews, "Quantifying Historical Carbon and Climate Debts among Nations," *Nature Climate Change* 6 (2016): 60–64.

[7] Sven Beckert, Bosma Ulbe, Mindi Schneider, and Eric Vanhaute, "Commodity Frontiers and the Transformation of the Global Countryside: A Research Agenda," *Journal of Global History* 16, no. 3 (2021): 435–450. For a history of the world economy centered on the exploitation of cheap resources, see Raj Patel and J. W. Moore, *A History of the*

considering the socio-ecological parameters and legacies of Eurasian growth require a lexicon that transcends established market/plan and capitalism/socialism dichotomies.[8] The critical scholar Andreas Malm introduced the term *fossil economy* to emphasize the link between capitalism, fossil fuels, and carbon emissions.[9] This book employs the notion of the fossil economy to place the case of peat in central Russia within a larger narrative of social and environmental change driven by the spiraling energy demand of an expanding industrial metabolism. Fossil economy also allows us to think through the continuities across the shift from late imperial capitalism to Soviet state socialism and to make sense of the parallels between Soviet state socialism and Western capitalism that permeate the history of peat despite its many particularities. Russia's modern economy, rather than being an anomaly rooted in political institutions and ideology, fits squarely within a global history of economic expansion and associated socio-ecological changes that reverberate locally and at a planetary level. Envisioned as the main contender of capitalism in the twentieth century, the Soviet economy never overcame but reinforced a productivist economic logic that caused a metabolic rift and continues to undermine the foundations of life on earth.[10]

Two premises guide this book. First, any productive activity relies on multiple forms of ecological and social reproduction, which conventional notions of the economy and economic history tend to obscure.[11] Second, as the social and material process that renders nature consumable, labor is

World in Seven Cheap Things: A Guide to Capitalism, Nature, and the Future of the Planet (London: Verso Books, 2018).

8 Alessandro Stanziani has conceptualized Eurasian growth and situated it within global history. His valuable narrative does not discuss the ecological premises of growth, however, nor does it extend into the Soviet period. See Alessandro Stanziani, *After Oriental Despotism: Eurasian Growth in a Global Perspective* (London: Bloomsbury Academic, 2014).

9 On fossil economy as a heuristic lens, see Andreas Malm, *Fossil Capital: The Rise of Steam Power and the Roots of Global Warming* (Brooklyn, NY: Verso Books, 2016), 11–15.

10 The literature on the metabolic rift and its origins in Marx's critique of political economy is extensive. See, for instance, Kohei Saito, *Marx in the Anthropocene: Towards the Idea of Degrowth Communism* (Cambridge: Cambridge University Press, 2022); John Bellamy Foster, "Marx's Theory of Metabolic Rift: Classical Foundations for Environmental Sociology," *American Journal of Sociology* 105, no. 2 (1999): 366–405.

11 On this point, see Stefania Barca, *Forces of Reproduction: Notes for a Counter-Hegemonic Anthropocene* (Cambridge: Cambridge University Press, 2020); Marina Fischer-Kowalski and Helmut Haberl, "Social Metabolism: A Metrics for Biophysical Growth and Degrowth," in Joan Martínez-Alier and Roldan Muradian, eds., *Handbook of Ecological Economics* (Cheltenham, UK: Edward Elgar, 2015), 100–138; Nancy

an essential category for understanding how economic growth in the past has relied on and affected both humans and the nonhuman world.[12] Erected in 1989, the monument of the female peat worker in front of Shatura's local museum appears like a distant echo of the early Soviet cult of labor, a celebration of workers' contribution to the building of socialism. Despite its naivety, the monument points to a crucial aspect of the history told in this book: the hard physical labor through which workers turned the matter stored in peatlands into fuel which, once converted into heat or electricity, powered machines and illuminated factories, streets, and homes. From the imperial period until the 1950s, this labor came primarily from seasonal, often female, workers who migrated between farm work and peat extraction every year. Production and reproduction were closely entangled in workers' lives. Even when seasonal labor declined in the late Soviet period, workers retained their dual role as employees of the peat industry and food producers, as they cultivated garden plots in their settlements. The peat industry provided cheap fuel to manufacturers and power plants and gave an income to peat workers. But it also took a toll. Extraction degraded wetlands and undermined their ability to accumulate peat as workers carried the marks of extractive labor on their bodies. In telling their history, this book examines not only how Russia's fossil economy placed the burden of its energy demands on people and nature at its margins but also how workers and peatlands responded to, accommodated, and resisted the demands imposed on them.

This effort gains a lot from thinking about the past in more-than-human terms. The environmental historians Emily O'Gorman and Andrea Gaynor have called for histories "that are attentive to the perpetually changing set of social, symbolic, ontological, and material relations through which historical actors – human and nonhuman – are co-constituted."[13] This book highlights the vitality of the material

Fraser, "Climates of Capital: For a Trans-Environmental Eco-Socialism," *New Left Review* no. 127 (2021): 94–127.

[12] Stefania Barca, "Labouring the Earth: Transnational Reflections on the Environmental History of Work," *Environment and History* 19, no. 1 (2014): 3–27; Simon Schaupp, *Stoffwechselpolitik: Arbeit, Natur und die Zukunft des Planeten* (Berlin: Suhrkamp, 2024).

[13] Emily O'Gorman and Andrea Gaynor, "More-Than-Human Histories," *Environmental History* 25, no. 4 (2020): 711–735, 717. For a geographical perspective, see Sarah Whatmore, "Materialist Returns: Practising Cultural Geography in and for a More-Than-Human World," *Cultural Geographies* 13, no. 4 (2006): 600–609.

world and its role in shaping the human experience at individual and societal levels. Peatlands and peat were not simply there; they constantly changed because of how humans interacted with and acted upon them and because of influences beyond human control. The amount of sunshine or rain, soil and air temperatures, the growth of plants, the spread of malaria-bearing mosquitoes, which made thousands of workers sick, and fire – all these factors played their part in making the history that *Burning Swamps* seeks to unearth. "More often than not," the historian Timothy McCain wrote, "it is not so much the case that abstract human minds interact with abstract sociocultural phenomena but rather that embodied human beings interact with material organisms and things that have lives, histories, and trajectories of their own."[14] Contemplating entanglements across the boundaries between humans and other species and between the living and the inert does not overwrite more familiar histories of social inequality and the uneven distribution of power. Instead, it reveals how social categories such as gender, race, and class are always embedded, reproduced, and contested within relationships that extend beyond the human world.[15] The peat fuel that supplied machines and electric power plants in central Russia was a product of both the often unruly interactions between humans and peatlands, as well as the social hierarchies that enabled the peat industry's appropriation of cheap rural labor, without which most peat would have stayed in the ground.

AN AWKWARD MATTER AND ITS HISTORICAL SIGNIFICANCE

Peat is an awkward matter that unsettles the categories of current debates surrounding energy and climate change. Peat accumulates in wetlands, where waterlogged conditions delay the breakdown of trees, mosses, and

[14] Timothy J. LeCain, *The Matter of History: How Things Create the Past* (Cambridge: Cambridge University Press, 2017), 14.

[15] For a critique of new materialist scholarship that takes social relations seriously without denying the agency of matter, see Scott Kirsch and Don Mitchell, "The Nature of Things: Dead Labor, Nonhuman Actors, and the Persistence of Marxism," *Antipode: A Radical Journal of Geography* 36, no. 4 (2004): 687–705. For a perspective from environmental history, see Gregg Mitman, "Living in a Material World," *The Journal of American History* 100, no. 1 (2013): 128–130.

other plant litter, leading to the build-up of organic matter at a rate of often no more than one millimeter per year. During this process, massive amounts of carbon are sequestered in the ground, making peatlands the most significant terrestrial carbon stores and sinks on the planet. The amount of carbon dioxide they hold per unit of land greatly exceeds that of forests.[16] Unlike geographically concentrated minerals, peat stretches across large territories close to the earth's surface. Extraction can be simple, sometimes requiring no more than a spade. However, referring to this activity as "harvesting" is inaccurate because extraction disrupts peat formation, particularly when accompanied by intensive drainage. Since most peat layers extend just a few meters into the ground, the pace of extraction may easily surpass the material build-up and severely reduce or deplete the peat layer in a given place.[17] Reflecting peat's peculiar state at the intersection of becoming and being, its categorization as a resource has caused considerable debate. Although peat is considered fossil from a phenomenological perspective – in the sense of being "conserved by burial," as the peatland scientist Hans Joosten described it – and its burning results in carbon dioxide emissions higher than those of anthracite or oil, industrial lobby groups have pushed to classify it as "renewable" or "biomass" to justify further extraction.[18] The IPCC does not assign peat to either biomass or fossil fuels, though countries report carbon dioxide emissions from burning peat under the fossil fuel category because of peat's high emission factors.[19]

These conceptual dilemmas resonate in the field of energy history, where peat straddles the distinction between "traditional" and "modern" or "organic" and "inorganic" (mineral) fuels.[20] Peat seems

[16] E. Maltby and M. C. F. Proctor, "Peatlands: Their Nature and Role in the Biosphere," in Eino Lappalainen, ed., *Global Peat Resources* (Jyskä: International Peat Society, 1996), 11–19, 12; Hans Joosten and Donal Clarke, *Wise Use of Mires and Peatlands: Background and Principles including a Framework for Decision-Making* (Devon: International Mire Conservation Group & International Peat Society, 2002), 25.

[17] For a general introduction to peat as a fuel and its global significance, see World Energy Council, *World Energy Resources: 2013 Survey* (London: World Energy Council, 2013), chapter 6.

[18] Hans Joosten, "The International Peat Society: Fossil or Renewable? An Analysis of the IPS Stand towards Peat Renewability and Climate Change," *IMCG Newsletter* no. 2 (2007): 4–19, 6.

[19] IPCC, *2006 IPCC Guidelines for National Greenhouse Gas Inventories*, Vol. 2, Energy (Hayama: IGES, 2006), chapter 1, 1.15, fn 5.

[20] Reflecting the limitations of these distinctions, a recent survey of European energy history includes peat in a short section on "new fuels" within a subchapter on "traditional sources" and then again in a subchapter on "The transition to coal." See Astrid

to have a natural place in pre-modern times, for which its use as a fuel is well documented.[21] However, it is strikingly absent from depictions of industrial economies and their energy sources. Apart from Alexander Etkind's cultural history of natural resources, which points to peat's often overlooked significance in global energy history, most surveys mention peat only in passing or ignore it altogether.[22] The "adolescent fossil fuel," as the environmental historian Alfred W. Crosby once called it, has a much lower energy density than most types of coal, oil, and natural gas. As a result, the conventional narrative of the fossil fuel age as an epoch in which highly potent fuels, mostly of fossil origin, have replaced organic and inefficient energy sources struggles to account for peat. The Dutch reliance on peat fuel during the early modern period tends to be treated as a historical exception during a transitional moment before coal, the "real" fossil fuel, prevailed.[23] In the words of the historian E. A. Wrigley, "peat can only provide an escape from the normal energy problems of an organic economy for a relatively brief period of time."[24]

Narratives focused on transition risk underestimating the biophysical weight and the diverse social experiences linked to different historical

Kander, Paolo Malanima and Paul Warde, *Power to the People: Energy in Europe over the Last Five Centuries* (Princeton: Princeton University Press, 2013), 60–62; 109–110.

[21] Petra J. E. M. van Dam, "Sinking Peat Bogs: Environmental Change in Holland, 1350–1550," *Environmental History* 6, no. 1 (2001): 32–45; John Feehan and Grace O'Donovan, *The Bogs of Ireland: An Introduction to the Natural, Cultural and Industrial Heritage of Irish Peatlands* (Dublin: University College Dublin Press, 1996), 3–11; Hans Joosten, "Human Impacts: Farming, Fire, Forestry and Fuel," in Edward Maltby and Tom Barker, eds., *The Wetlands Handbook* (Chichester: Wiley-Blackwell, 2009), 689–718, 706; William N. TeBrake, *Medieval Frontier: Culture and Ecology in Rijnland* (College Station: Texas A&M University Press, 1985), 234–238; Paul Warde and Tom Williamson, "Fuel Supply and Agriculture in post-Medieval England," *The Agricultural History Review* 62, no. 1 (2014): 61–82, 63–67.

[22] Alexander Etkind, *Nature's Evil: A Cultural History of Natural Resources*. Translated by Sara Jolly (Cambridge: Polity Press, 2021), 231–237.

[23] Alfred W. Crosby, *Children of the Sun: A History of Humanity's Unappeasable Appetite for Energy* (New York: W. W. Norton, 2006), 61. Apart from this mention, peat does not feature any further in Crosby's book. Vaclav Smil's prominent energy history survey equally mentions peat only in passing. Vaclav Smil, *Energy and Civilisation: A History* (Cambridge, MA: MIT Press, 2017), 196, 228, 421.

[24] E. A. Wrigley, *Energy and the English Industrial Revolution* (Cambridge: Cambridge University Press, 2010), 23, 221. An important quantitative study on the diverse nature of energy consumption in England and Wales since the early modern period does not account for peat, though it admits it "may have been of local importance." See Paul Warde, *Energy Consumption in England and Wales, 1560–2000* (Naples: Naples: Consiglio Nazionale delle Ricerche, 2007), 22.

energy regimes. The historians Christophe Bonneuil and Jean-Baptiste Fressoz argue that energy history should be narrated as a series of additions if it were to understand the material impact of past energy use. Instead of moving along a linear path from wood to coal, then oil and natural gas, and later nuclear energy, industrial economies consumed many kinds of energy simultaneously within a continuous upward spiral of resource use.[25] Blurring the distinction between "traditional" and "modern" energy sources, this framework allows us to acknowledge histories of peat as part of this trend instead of dismissing them as irrelevant or abnormal. Crucially, peat consumption in the Netherlands peaked between the late nineteenth and early twentieth century – and not in the early modern period with which it is usually associated.[26] In the nineteenth century, peat fuel was used in small industries and sometimes even railway transport in Canada, Denmark, France, Ireland, and Germany, though there are likely many more examples. As a rule, this practice was regionally concentrated and declined in the first half of the twentieth century. Fuel insecurity during the World Wars could revive local interest in peat.[27] In Ireland, with its strong but gradually fading tradition of private peat cutting, decentralized, often illegal extraction regained importance amidst energy price hikes following Russia's full-scale invasion of Ukraine in 2022.[28]

[25] Christophe Bonneuil and Jean-Baptiste Fressoz, *The Shock of the Anthropocene: The Earth, History and Us*. Translated by David Fernbach (London: Verso Books, 2016), 100–104. For a quantitative perspective, see Richard York and Shannon Elizabeth Bell, "Energy Transitions or Additions? Why a Transition from Fossil Fuels Requires More than the Growth of Renewable Energy," *Energy Research & Social Science* no. 51 (2019): 40–43.

[26] Kander et al., *Power to the People*, 62.

[27] There is no systematic comparative study on this subject. For the individual cases mentioned, see Sylvain Dournel, "De l'exploitation historique de la tourbe dans la vallée de la Somme à la mise en nature des zones humides dans l'agglomération amiénoise: Mutations paysagères et sociales," in J. M. Derex and F. Grégoire, eds., *Histoire économique et sociale de la tourbe et des tourbières* (Cordemais: Æesturia & GHZH, 2009), 211–228; Corinna Malek, "Torfenergie und ihre Folgen in Bayern," in *Kulturerbe Energie: Zeugnisse der Energiegewinnung und Energienutzung als Kulturlandschaftselemente entdecken* (Bonn: Bund Heimat und Umwelt, 2015), 54–64; Mogens Schlüter, "The Use of Peat in Danish Glassworks, 1825–1945," *Journal of Glass Studies* 30 (1988): 94–101; Richard W. Unger and John Thistle, *Energy Consumption in Canada in the 19th and 20th Centuries* (Naples: Consiglio Nazionale delle Ricerche, 2013), 60–62; Barry G. Warner and Pierre Buteau, "The Early Peat Industry in Canada, 1864–1945," *Geoscience Canada* 27, no. 2 (2000): 57–66.

[28] Rory Carroll, "'Like an Oilwell in Your Back Yard': Irish People Turn to Cutting Peat to Save on Energy Bills," *The Guardian* (December 12, 2022), www.theguardian.com/

However, peat remained relevant beyond the private household or individual enterprises, where energy infrastructure was designed around it. This is evident in Ireland, whose peat-fueled power stations operated until the early twenty-first century, as well as in Finland and Sweden, where several central heating power plants and district heating that had fully or partially relied on peat are now in the process of being converted to other energy sources. The oil crises of the 1970s helped to retain and even strengthen peat's role in all three cases.[29] In the Baltic republics, peat-based heat and power systems originating from the Soviet period have similarly survived into the new millennium, even though their reliance on the resource is declining.[30] Between 2013 and 2017, nearly 65 percent of all peat extracted in the European Union went into energy, while the remaining served nonenergy (mostly horticultural) purposes. During this period, Finland and Ireland were the leading consumers of peat fuel.[31] While its role as an energy carrier is rapidly shrinking, peat has had a much longer and more diverse history than most energy historians would admit. It certainly mattered beyond domestic consumption and the early stages of industrial growth. As this book demonstrates, the case of Russia is a crucial chapter in the history of how the energy hunger of expanding industrial economies degraded the world's peatlands.

THE NATURE OF RUSSIA'S FOSSIL ECONOMY

Peat has been neglected in Eurasian energy histories no less than in the broader field of energy history. Reflecting Russia's longstanding role as an

world/2022/dec/12/like-an-oilwell-in-your-back-yard-irish-turn-to-cutting-peat-to-save-on-energy-bills. Accessed July 13, 2024.

29 Lily Toomey, "'A National Resource of Primary Importance': Transforming Ireland's Peatlands in the Twentieth Century," PhD Dissertation, Trinity College Dublin (2025); Nima Javanshir et al., "Abandoning Peat in a City District Heat System with Wind Power, Heat Pumps, and Heat Storage," *Energy Reports* 8 (2022): 3051–3062; Jan Kunnas and Timo Myllyntaus, "Postponed Leap in Carbon Dioxide Emissions: The Impact of Energy Efficiency, Fuel Choices and Industrial Structure on the Finnish Economy, 1800–2005," *Global Environment* 2, no. 3 (2009): 154–189, 177; Lorenzo Di Lucia and Karin Ericsson, "Low-Carbon District Heating in Sweden – Examining a Successful Energy Transition," *Energy Research & Social Science* 4 (2014): 10–20.

30 World Energy Council, *World Energy Resources*, 268–269.

31 This number is based on data for the EU27 presented in Oliver Hirschler and Bernhard Osterburg, "Peat Extraction, Trade and Use in Europe: A Material Flow Analysis," *Mires and Peat* 28, no. 24 (2022), table 1, doi: 10.19189/MaP.2021.SNPG.StA.2315.

energy power, the field has focused on the geopolitics of fossil fuels, a topic that has moved to the fore of public attention since Russia's full-scale invasion of Ukraine in 2022. As the political scientist Margarita M. Balmaceda demonstrated, participation in "Russian energy chains," which for decades linked fuel deposits in Siberia to commercial, public, and private consumers in central and western Europe, profoundly impacted domestic and foreign politics all along the value chain – from Russia, through "midstream" countries such as Ukraine and Belarus, to the importing nations in the EU.[32] Historians have traced these ties back to the Cold War period, showing how fossil fuel exports allowed the Soviet Union – and later the Russian Federation – to earn foreign currency and project power abroad, as countries at the receiving end benefited from cheap energy that fueled decades of economic growth.[33] Peat, which was deemed a "local fuel" (*mestnoe toplivo*) since imperial times and never attained the status of an international commodity, does not fit into this picture. According to the economist Lev B. Kafengauz, peat accounted for only 3.3 percent of the overall fuel consumption in the Russian Empire in 1900.[34] In 1960, peat fuel represented no more than 2.9 percent of all fuel produced in the Soviet Union. Its share declined to a mere 0.4 percent in 1980, making it the least significant fuel alongside firewood and shale.[35] From a statistical point of view, ignoring peat might seem justifiable.

But where do these narratives and figures leave a place like Shatura, the women I listened to, and central Russia's degraded peatlands? Aggregated statistics may fail to identify historical importance and render regional

[32] Margarita M. Balmaceda, *Russian Energy Chains: The Remaking of Technopolitics from Siberia to Ukraine to the European Union* (New York: Columbia University Press, 2021).

[33] There is a rich scholarship on this topic. See Thane Gustafson, *The Bridge: Natural Gas in a Redivided Europe* (Cambridge, MA: Harvard University Press, 2020); Per Högselius, *Red Gas: Russia and the Origins of European Energy Dependence* (New York: Palgrave Macmillan, 2013); Jeronim Perović, *Rohstoffmacht Russland: Eine globale Energiegeschichte* (Cologne: Böhlau, 2022); Jeronim Perović, ed., *Cold War Energy: A Transnational History of Soviet Oil and Gas* (Cham: Palgrave Macmillan, 2017); Felix Rehschuh, *Aufstieg zur Energiemacht: Der sowjetische Weg ins Erdölzeitalter 1930er bis 1950er Jahre* (Cologne: Böhlau, 2018).

[34] L. B. Kafengauz, *Ėvoliutsiia promyshlennogo proizvodstva Rossii (posledniaia tret' XIX v.–30-e gody XX v.)* (Moscow: Ėpifaniia, 1994), 31.

[35] David Wilson, *The Demand for Energy in the Soviet Union* (London: Croom Helm, 1983), 12.

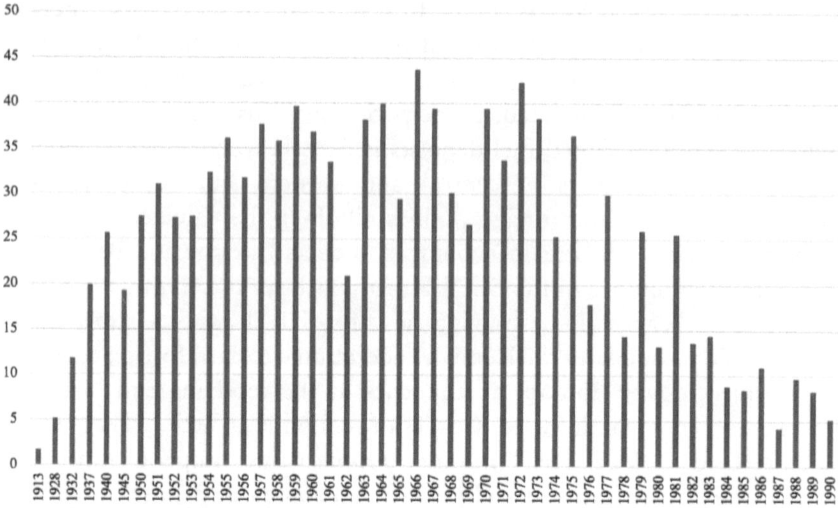

FIGURE I.3 Peat fuel production in the RSFSR (million tons) reflecting substantial annual variation and the overall rise and decline of the peat fuel industry in the twentieth century. Based on *Narodnoe Khoziaistvo RSFSR* (Moscow: Gosudarstvennoe statisticheskoe izdatel'stvo, 1956–1990). Note that peat extracted for agricultural uses, which in the late Soviet period surpassed the amount of peat used as a fuel, is not included here.

variety invisible. Focusing on absolute instead of relative numbers, on physical matter instead of percentages, we can draw a more accurate picture of the past, one that accommodates how peat extraction shaped human and nonhuman life at the margins of Russia's fossil economy. From the late nineteenth century, workers and machines removed ever-increasing volumes of peat from central Russia's wetlands, turning it into fuel that supplied factories and power utilities. Several of these utilities proved crucial for maintaining electricity supply in cities and industries during the Second World War. The weather could support or undermine these efforts, causing considerable variation in annual extraction. Still, a long-term perspective reveals a clear pattern. Peat extraction exhibited strong growth from the late imperial period and accelerated during Soviet times. Within the RSFSR, annual peat fuel production rose from just 1.7 million in 1913 to some 40 million metric tons in the late 1960s and early 1970s, after which it declined (Figure I.3). Though its contribution to the aggregated fuel balance was small, the absolute amount of peat extracted for fuel in the Soviet Union exceeded that of any other country in the world. In 1980, when peat fuel consumption was already shrinking, the Soviet Union still accounted for almost 90 percent of the globally produced peat fuel.[36]

This history had a distinct geography. Peatlands are among the most prominent features of the landscape in Northern Eurasia. Taken together, peatlands and so-called paludified (shallow-peat) land, with a peat layer of fewer than 30 centimeters, account for 20 percent of the territory within the borders of the Russian Federation. Peatlands span vast parts of the Permafrost zone in Siberia and the Far East and are also found in many areas west and northwest of the Ural mountain chain (Figure I.1, p. 1). Peat extraction for energy generation and agricultural products, which surpassed fuel peat in the late Soviet period, affected these lands unevenly. The most intense exploitation occurred in the Moscow region and adjacent parts of the Vladimir region, where industrial producers came to value peat in the second half of the nineteenth century. Peatlands in Ryazan, Tver, Yaroslavl, Nizhny Novgorod, and Saint Petersburg were severely marked by the peat industry, too, while wetlands in Siberia and the Far East have been affected to a much lesser extent. During Soviet times, large-scale extraction transformed peatlands in the Belarussian SSR, parts of Ukraine, and the Baltic republics. In the Baltics, the Soviet peat industry drew on an earlier history of extraction that went back to the nineteenth century.[37]

Reconstructing how peat was extracted and used as a fuel in manufacturing and the power industry, *Burning Swamps* provides a corrective to dominant narratives in Eurasian energy history. This field may not suffer from what the historian Christopher F. Jones called "petromyopia," an overemphasis on all things relating to oil at the expense of other energy

[36] Bord na Móna, *Fuel Peat in Developing Countries*. World Bank Technical Paper Number 41 (Washington DC: The World Bank, 1985), 8.

[37] Tatiana Minayeva and Andrey Sirin, eds., *A Quick Scan of Peatlands in Central and Eastern Europe* (Wageningen: Wetlands International, 2009), 26–29; A. A. Sirin et al., "Mapping of Peatlands in the Moscow Oblast Based on High-Resolution Remote Sensing Data," *Contemporary Problems of Ecology* 7, no. 1 (2014): 808–814, 809. See also the relevant country chapters in Hans Joosten, Franziska Tanneberger and Asbjørn Moen, eds., *Mires and Peatlands of Europe: Status, Distribution and Conservation* (Stuttgart: Schweizerbart Science Publishers, 2017). On the appropriation of peatlands in the Baltic region during the nineteenth century, see Esa Ruuskanen, "The Emergence of Baltic *Moorkultur*: Visions of Scientific–Technological Mastery of Peatlands in the Age of Great Social Change, 1850–1914," *History and Technology* 34, no. 3–4 (2018): 213–234.

sources.[38] Still, the tendency to think about energy primarily through the prism of international relations has created blind spots. Except for hydro- and nuclear power, the intertwined histories of energy, labor, and nature are poorly understood.[39] Several case studies with a regional focus, along with historical research on renewables, have recently opened up exciting avenues, highlighting shifting conceptualizations and uses of energy resources in Eurasia's recent past.[40] It is this conversation that this book seeks to advance. Tracing how the extraction and use of peat, a not very potent and supposedly unimportant fuel, transformed rural livelihoods and degraded central Russia's wetlands, I underscore the historical relevance of the statistical margins and the value of regional perspectives in modern energy history.

A HISTORY ACROSS SCALES

The history of peat in Russia's fossil economy operates across multiple temporal and geographical scales. Considering sources from various parts of European Russia, this book primarily focuses on the Meshchera Lowland, a landscape marked by extensive wetlands, forests, and lakes

38 Christopher F. Jones, "Petromyopia: Oil and the Energy Humanities," Humanities 5, no. 2 (2016), doi: https://doi.org/10.3390/h5020036.

39 Kate Brown, *Manual for Survival: A Chernobyl Guide to the Future* (London: Penguin, 2019); Klaus Gestwa, *Die Stalinschen Großbauten des Kommunismus: Sowjetische Technik- und Umweltgeschichte, 1948–1967* (Munich: Oldenbourg, 2010); Anna Olenenko, "'Novoe nashe more - novoe nashe gore': Konflikt mezhdu ukrainskim i sovetskim v bor'be za konstruirovanie landshafta Nizhnego Podneprov'ia," *Ab Imperio* no. 1 (2019): 125–152.

40 See Andy Bruno, *The Nature of Soviet Power: An Arctic Environmental History* (Cambridge: Cambridge University Press, 2016). Felix Frey explicitly mentions peat as a major blind spot in the scholarship. Felix Frey, *Arktischer Heizraum: Das Energiesystem Kola zwischen regionaler Autarkie und gesamtstaatlicher Verflechtung 1928–1974* (Cologne: Böhlau, 2019), 29. On the history of renewables, see Benjamin Beuerle, "A Dormant Giant: Renewable Energy in the Soviet Union and Russia (1970s–Present)," *Russian Review* 84, no. 2 (2025): 267–282 Daria Gritsenko and Hilma Salonen, "A Local Perspective on Renewable Energy Development in the Russian Artic," *Elementa: Science of the Anthropocene* 8, no. 1 (2020): 441, doi: https://doi.org/10.1525/elementa.441; Felix Frey, "Putting Oceans to Work: Tidal Energy in the USA and the USSR, 1930–1970," *History and Technology* 37 no. 4 (2021): 487–504; Natalia Nikiforova, "'Useless Fossils', Precious Waste, and Streams of Energy: Soviet Electrification and Natural Resources for the Socialist Future (1920s–1930s)," *Technology and Language* 4, no. 2 (2023): 72–87; Daniela Russ, "'Socialism Is Not Just Built for a Hundred Years': Renewable Energy and Planetary Thought in the Early Soviet Union (1917–1945)," *Contemporary European History* 31, no. 4 (2022): 491–508.

between the administrative regions of Moscow, Vladimir, and Ryazan. Peat extraction in Meshchera took off in the imperial period and gained pace under Soviet power, particularly in the second half of the twentieth century. From the early 1930s, it fell under the responsibility of the Shatura Peat Trust (Shaturtorf), an umbrella organization for over 10 peat companies which secured fuel for the Shatura Power Plant. Following the decline of the peat industry in the late Soviet period, parts of the region's remaining plant and animal life were protected. In 1994, the floodplains of the rivers Oka and Pra, a small Oka tributary, in southern Meshchera were designated Wetlands of International Importance under the 1971 Ramsar Convention. From the early 2000s, and particularly following the 2010 fires, several former peat extraction sites have been rewetted.[41]

Examining this area's intertwined social and environmental history through a century of extraction, *Burning Swamps* opens up new perspectives on the geography of marginality in Russia's empires. The stories told here highlight the historical significance of spaces between imperial metropoles and peripheries, spaces that evade familiar administrative and analytical distinctions between the city and the countryside.[42] At the same time, this book expands conversations about place and region in Russia's past into the material world.[43] Peat extraction and associated developments in transport and settlement left lasting imprints on physical environments in areas like Meshchera, as myths, stories, and people's lived experiences endowed the changing landscapes with cultural significance. This book understands these transformations as inherently interrelated. It illustrates that the making of places and regions in modern Russia was as much a cultural as a socioecological process which

[41] On the background, see Alan D. Roe, *Into Russian Nature: Tourism, Environmental Protection, and National Parks in the Twentieth Century* (New York: Oxford University Press, 2020), 227–231; "Oka & Pra River Floodplains," Ramsar site information: https://rsis.ramsar.org/ris/671?language=en. Accessed March 18, 2025. For case studies of selected Meshchera sites, see N. P. Akhmet'eva et al., *Torfianye bolota tsentral'nykh oblastei Evropeiskoi chasti Rossii: Ikh transformatsiia za poslednie 100 let* (Moscow: GEOS, 2020), 62–93.

[42] The anthropologist Jeremy Morris has taken a similar (though not environmental) approach for the post-Soviet period. Jeremy Morris, *Everyday Post-Socialism: Working-Class Communities in the Russian Margins* (London: Palgrave Macmillan, 2016).

[43] Fundamental for this discussion has been Catherine Evtuhov, *Portrait of a Russian Province: Economy, Society, and Civilization in Nineteenth-Century Nizhnii Novgorod* (Pittsburgh: University of Pittsburgh Press, 2011).

reconfigured physical and human geographies in ways that effectively blurred the distinction between them.[44]

Rooted in local particularities, the forgotten history of peat offers insights into Russia's place in broader histories of growth. Chronology provides important clues. The economic history of Russia's empires is typically organized around critical political turning points or economic reforms, such as the Stolypin reforms, the Bolshevik takeover in 1917, or Stalin's economic revolution. However, from the perspective of peatlands and peat workers, there was a lot of continuity across these moments. Significant changes occurred in the late 1950s when new machinery reduced the industry's labor dependence, expanded its spatial reach, and amplified the environmental damage it inflicted. The large-scale exploitation of peatlands for industries and power plants might seem like an outlier within the master narrative of the fossil fuel age. Interpreting it as evidence of Eurasia's alleged economic *sonderweg* would be misleading though. Instead, we should understand it as one of the many manifestations of the Great Acceleration, marked by unprecedented resource use, growth, and ecological disruption following the Second World War.[45] The history of peat extraction also muddies clear-cut distinctions between both the imperial and the Soviet economy on the one hand and the Soviet economy and market economies on the other. Indeed, as the historian Bathsheba Demuth argued, market and plan were not so different in their disregard for nonhuman time and the rhythms of natural reproduction.[46] When viewed through frameworks that move beyond the chronologies of political history and emphasize the socioecological premises of economic growth, the history of peat confirms global trends rather than diverging from them.

The margins of Russia's fossil economy also invite us to view the past from a planetary perspective in ways that – other than universalizing notions of humanity as a singular force of ecological decline –

[44] For a recent collection of case studies addressing issues of place in Eurasian history, see David Moon, Nicholas B. Breyfogle and Alexandra Bekasova, eds., *Place and Nature: Essays in Russian Environmental History* (Cambridgeshire: White Horse Press, 2021).

[45] John R. McNeill and Peter Engelke, *The Great Acceleration: An Environmental History of the Anthropocene since 1945* (Cambridge, MA: Belknap Press of Harvard University Press, 2014).

[46] Bathsheba Demuth, *Floating Coast: An Environmental History of the Bering Strait* (New York: W. W. Norton, 2019).

emphasize accountability alongside the contingent histories and local manifestations of the current environmental predicament.[47] Formed as a result of the Holocene glacial retreat, peatlands in the northern Hemisphere have sequestered carbon over thousands of years – time spans that most historians would consider outside of their domain. However, drainage and extraction turned many of them from carbon sinks into carbon emitters, illustrating that, as the anthropologist Richard D. G. Irvine wrote, "life is lived in relation to the geology, not just on top of it, and that through these interactions deep time protrudes into everyday life."[48] Scientists consider peatland degradation a significant factor in human-caused greenhouse gas emissions since the nineteenth century.[49] Peat fires, more likely to happen when the water table is low, exacerbate the positive feedback loop between degraded peatlands and climate change.[50] These issues are particularly evident when deeply drained peatlands are abandoned after extraction, a common occurrence in many parts of Russia as the state retreated from the countryside following the Soviet collapse.[51] Far from being a minor episode, the forgotten history of peat in Russia's fossil economy is both part of the global trajectory of wetland destruction in the modern era and a process deeply reflective of local circumstances. One of the

[47] Such an approach is advocated by Charlotte Wrigley, *Earth, Ice, Bone, Blood: Permafrost and Extinction in the Russian Arctic* (Minneapolis: University of Minnesota Press, 2023). On the planetary as a historical category, see Dipesh Chakrabarty, *The Climate of History in a Planetary Age* (Chicago: University of Chicago Press, 2021). For a nuanced discussion of the concept of the Anthropocene and the problematic notion of a singular human species, see Rob Nixon, "The Anthropocene: The Promise and Pitfalls of an Epochal Idea," in Gregg Mitman, Marco Armiero and Robert S. Emmett, eds., *Future Remains: A Cabinet of Curiosities for the Anthropocene* (Chicago: The University of Chicago Press, 2018), 1–18.

[48] Richard G. D. Irvine, *An Anthropology of Deep Time: Geological Temporality and Social Life* (Cambridge: Cambridge University Press, 2020), 38.

[49] Jens Leifeld et al., "Intact and Managed Peatland Soils as a Source and Sink of GHGs from 1850 to 2100," *Nature Climate Change* 9 (2019): 945–947.

[50] Merritt R. Turetsky et al., "Global Vulnerability of Peatlands to Fire and Carbon Loss," *Nature Geoscience* 8, no. 1 (2015): 11–14; S. L. Wilkinson et al., "Wildfire and Degradation Accelerate Northern Peatland Carbon Release," *Nature Climate Change* 13 (2023): 456–461.

[51] J. M. Waddington et al., "Cutover Peatlands: A Persistent Source of Atmospheric CO_2," *Global Biogeochemical Cycles* 16, no. 1 (2001): 1–7, doi: https://doi.org/10.1029/2001GB001398; A. Sirin et al., "Addressing Peatland Rewetting in Russian Federation Climate Reporting," *Land* 10, no. 11, 1200 (2021): 1–17, 2–3, doi: https://doi.org/10.3390/land10111200; A. Vozbrannaya et al., "After Wildfires and Rewetting: Results of 15+ Years' Monitoring of Vegetation and Environmental Factors in Cutover Peatland," *Diversity* 15, no. 3 (2023), doi: https://doi.org/10.3390/d15010003.

countless chapters in the history of our planetary emergency, the case of peat straddles the boundaries between here and there, now and then, calling us to think about human history beyond the human.

NATURE'S WEALTH AND THE INTRICACIES OF LANGUAGE

Ecologists have developed a nuanced vocabulary for the various ecosystems that evade the land/water binary, though terminologies vary significantly across countries and languages, as do the natural formations they want to categorize. "Wetlands," a term that gained legal and political significance following the (Ramsar) Convention on the Protection of Wetlands of International Importance in 1971, is widely accepted as an umbrella category for environments that "occur wherever water meets land," as the leading NGO Wetlands International puts it.[52] The term encompasses diverse ecosystems such as salt marshes or river deltas and environments like rice paddies or salt pans that have been created by humans.[53] Peatlands, a subcategory of wetlands defined by a layer of naturally accumulated peat, are the most widespread type of wetlands in the world. There are different national conventions to differentiate various types of peatlands. Scientific publications in English use "mire" to refer to a peatland where peat formation is ongoing. Another important distinction in this literature is that between "fens," which are typically located in depressions and receive most of their water from the ground, and "bogs," whose water comes from precipitation.[54]

The Russian language has its own tradition for naming and classifying waterlogged lands. Most historical sources that inform this book used the word *boloto* (plural *bolota*), which frequently, but not always, carried negative connotations reflecting people's annoyance with or fear of environments they found difficult to understand or navigate. *Boloto* can mean many things in English, including swamp, marsh, peatland, wetland, fen, bog, mire, or quagmire. Sometimes, historical actors used more specific terms, such as *torfianoe boloto* ("peaty swamp"), *torfianik*

[52] www.wetlands.org/wetlands/. Accessed July 15, 2024.

[53] Broader definitions even include open water bodies, such as rivers and lakes. See the definition on the (Ramsar) Convention on Wetlands, at www.ramsar.org/about/our-mission/importance-wetlands. Accessed July 15, 2024.

[54] For an overview, see Hans Joosten et al., "Mire and Peatland Terms and Definitions in Europe," in Joosten et al., eds., *Mires and Peatlands of Europe*, 65–74.

("peatland"), or *torfianoe mestorozhdenie* ("peat deposit").[55] However, the simpler term *bolota* dominated and remained influential in the Russian discourse on peat and peatlands throughout the history covered in the following chapters. In the wake of the Ramsar Convention, the term "wetlands" was translated into Russian as *vodno-bolotnye ugod'ia* (literally "watery-marshy grounds"). In contrast to the English term, which has been widely adopted by scientists, politicians, and activists, its rather cumbersome Russian equivalent has not gained much public or political traction.

These linguistic intricacies resonate throughout this book. Since the narrative primarily focuses on peat and peat extraction, "peatland" and "wetland" are used interchangeably as descriptive terms. When translating or discussing primary sources, I use "swamp" to maintain a source's pejorative association with a place or landscape referred to as *boloto*; "peatland" and "mire" when the source's use of *boloto* aligns with the respective scientific definitions; or "wetland" when it is unclear whether the historical actors considered peat a defining trait of the marshy land they described.

* * *

Peat is not naturally a resource. Part I *Promising Environments* explores the mental and material process through which peat came to be seen and used as an energy carrier. Chapter 1 traces how, beginning in the eighteenth century, imperial elites projected visions of improvement and abundance onto Russia's wetlands, reimagining them as fuel deposits. It argues that peat extraction was primarily an elite project that imposed the developmentalist visions of the imperial state and industrial producers on peatlands and the people living with them. While most peasants continued valuing peatlands for what they offered above ground, elite groups conceptualized peat as a distinct substance rather than a component of complex wetland environments. Chapter 2 examines how peat became embedded in Russia's industrial metabolism. In central Russia, peat gained importance as an industrial fuel from the late nineteenth

55 On Russian peatland terminology and classification systems, see Andrey Sirin et al., "Russian Federation," in Hans Joosten, Franziska Tanneberger and Asbjørn Moen, eds., *Mires and Peatlands of Europe: Status, Distribution and Conservation* (Stuttgart: Schweizerbart Science Publishers, 2017), 591–616, 591–601.

century, inspiring technical elites to consider it a source of electric power. This idea was subsequently incorporated into the famous GOĖLRO-plan for the Electrification of Russia, which firmly anchored peat in the Soviet power industry. The early Soviet energy system, with its emphasis on regionally available energy sources, was not solely a product of Bolshevik power. Instead, it must be situated within longer trajectories of regionalized fuel use and the experience of a war-related fuel crisis that predated the 1917 Revolution.

Extraction sites and their surroundings are the focus of Part II *Working Environments*, which explores how the peat industry shaped workers' lived experiences and refashioned central Russia's wetland environments. Chapter 3 discusses the seasonal nature and gendered organization of labor. It reveals that, as an embodied, more than-human activity, peat extraction was an experience marked by social inequality and difference as well as by the uncertain material environments of extraction sites, where the weather, dysfunctional technology, and the physical interaction with peat caused injuries and accidents. But extraction also helped to make central Russia's peatlands more habitable. Looking at the physical transformations of the land, housing, and the changing cultural significance of peat extraction sites and their vicinities, Chapter 4 foregrounds the often overlooked role of workers' settlements as spaces of reproduction in the history of Russia's economy. Peat, this chapter contends, was not just a fuel but also a source for place-based feelings of belonging that allowed workers to embrace the margins of Russia's fossil economy as their home.

Part III *Unsettling Environments* discusses hazards, loss, and precarity as central aspects in the history of peat extraction. Focusing on malaria outbreaks and fire events, Chapter 5 highlights the agency of peatlands in the history of Russia's fossil economy. The chapter identifies peat extraction sites as spaces of environmental injustice and points to a crucial irony running through the history of human–peatland relationships in imperial and Soviet Russia: Peatlands had long been imagined as dangerous and useless, but they only turned into unsettling landscapes once they became part of Russia's industrial metabolism. Chapter 6 demonstrates that the decline of the peat industry in the late Soviet period, driven by the country's shift to Siberian fossil fuels and a rising scientific and cultural appreciation for intact wetlands, brought new challenges. The Soviet collapse made the end of extraction an unsettling experience. Many regions were cut off from the resources and services that had sustained them in previous decades, leading to social marginalization and massive fire events. This final chapter demonstrates that, instead of recovery, the end of extraction brought new forms of precarity. Peat's role as a fuel may have shrunk dramatically in the past decades, but the legacies of its extraction and use are bound to remain.

PART I

PROMISING ENVIRONMENTS: MATERIAL PREMISES OF GROWTH AND POWER

1

Appropriation

In 1910, peasant Andrei Efimov approached the regional branch of the Ministry of Agriculture to enquire whether he could rent a wetland close to his native village of Iarnevo in the province of Tver. Writing from the city of Saint Petersburg, where he presumably engaged in seasonal factory labor, the peasant hoped to use the plot to make hay or graze his livestock. Efimov argued that the lease would carry clear financial benefits for the state. Instead of maintaining a woodman to oversee a wet and fallow piece of land, the treasury could receive a regular income, while Efimov himself would invest his labor and take the risk of draining the area. Well aware that such a deal required that the monetary value of the plot be established, Efimov hurried to express his confidence in the administration's expertise to assess the land: "I cannot suggest a concrete price because it is uncertain which use there will be, as this is a swamp. It will be easier for the treasury ... to set the price."[1]

For Efimov, this was probably a straightforward suggestion. The matter looked much more complicated from the point of view of the administration though. The peasant's letter ended up in the Turf Master's Division, a small unit within the Department of Land Improvement tasked with the profitable management of state-owned wetlands. The staff there hesitated to accommodate Efimov's request because they lacked even the most basic sense of the piece of land that had caught the peasant's interest: Was there any peat in the swamp? And if this were the case, would Efimov's proposed plan damage the peat,

[1] Copy of Efimov's letter (1910), RGIA f. 425, op. 1, d. 35, l. 127.

which could potentially be used as a source of fuel? After all, wouldn't the financial benefits for the treasury be much more substantial if the land was leased to someone interested in peat extraction rather than to a peasant interested in expanding animal husbandry? Given so many open questions, the officials were unwilling to consent: The plot had not yet been surveyed, so a decision would have to be postponed.[2]

The archives are silent about whether or not the administration followed up on Efimov's request. We don't know if the peasant eventually got to lease the land. Still, this anecdote is significant in that it highlights the different ways of seeing wetlands in late imperial Russia, the importance of social hierarchies in determining how they would be used, and the practical difficulties of turning wetlands into profitable use. Efimov's proposal aligned with the imperial administration's ambition to develop the farming sector and with the growing recognition of land improvement (*melioratsiia*) as a means to facilitate this goal.[3] Efimov's suggestion also fit with the administration's strategy of augmenting the state budget by letting private users pay for the temporary right to exploit state-owned assets, such as land, water, forests, or wetlands. And yet, the vague possibility that the plot could be turned into a peat extraction site prevented the administration from approving the peasant's request – apparently because the officials expected that the state would achieve higher lease payments from a hypothetical industrial entrepreneur and perhaps also because they regarded the extraction of peat to be a more timely way to use the land than letting animals graze on it. What constituted the best way to use Russia's wetlands was a matter of perspective, and the lack of information made it even more difficult for the officials to decide.

But why would state officials consider the wetland in Tver as a potential extraction site in the first place? After all, most peasants at the time appreciated such environments in horizontal terms: as spaces where they could hunt and gather berries, mushrooms, and other useful plants, as grazing grounds for their animals, or as potential farmland. When the officials speculated about the wetland as a reservoir of resources stored underneath the surface, they conceived of the land vertically, engaging in a mental operation which reduced peatlands to

[2] Internal correspondence about Efimov's request, ibid., ll. 126–129.

[3] Katja Bruisch, "The State in the Swamps: Territorialization and Ecosystem Engineering in the Western Provinces of the Late Russian Empire," *Zeitschrift für Ostmitteleuropa-Forschung* 68, no. 3 (2019): 345–368.

a single one of their multiple material components: peat. This abstraction may have contrasted with the rural population's perceptions and uses of wetlands, but it did not appear out of nowhere. That civil servants working in an office in Saint Petersburg were able to project such ideas onto a distant piece of land was the result of a long resource-making process which originated in the eighteenth century and gained traction with the dynamic development of industries during the late imperial period.

Peat extraction in Russia developed, by and large, as an elite project that imposed the interests of the central state and industrialists onto peatlands and the people living with them. Following scientists, noble landowners, entrepreneurs, and state officials, as they explored, mapped, and valued what they called "swamps" (*bolota*), this chapter charts how wetlands entered the elites' imagination as deposits of subterranean wealth and how imperial approaches to resources carried over into the Soviet period. The mental and, as time went on, the physical appropriation of wetlands as fuel deposits was not a straightforward process though. Peat sparked commercial ambition and scientific curiosity but also evoked suspicion and rejection. Plans to transform peatlands into extraction sites faced multiple drawbacks that reflected the prevalence of peaty soils in the Russian Empire, the technical difficulty of surveying these lands, and the initial reluctance to adopt the new resource. The messy and nonlinear process by which peatlands were appropriated as fuel deposits hints at the complicated interactions between matter and mind, culture and nature within the history of natural resources. Shaped by different ideas, interests, and actors as much as by the complex nature of the wetlands themselves, the case of peat underscores how misleading it would be to think about resources as static entities in Russian history. Instead, we should understand resources as historically contingent and constantly changing reflections of humans' mental and material interactions with the nonhuman world.[4] Even though peatlands are a central feature of the Eurasian environment today as much as when they first appeared on the radar of imperial elites, their meaning and their role in Russia's economy have not been fixed.

[4] Gavin Bridge, "Material Worlds: Natural Resources, Resource Geography and the Material Economy," *Geography Compass* 3, no. 3 (2009): 1217–1244; Etkind, *Nature's Evil*, 5; Tanya Richardson and Gisa Weszkalnys, "Introduction: Resource Materialities," *Anthropological Quarterly* 87, no. 1 (2014): 5–30.

RESOURCE-MAKING AS AN ELITE PROJECT

Considering the age and extent of Eurasia's peatlands, their appropriation as resources is a relatively recent phenomenon. The idea to systematically drain and exploit wetlands gained traction in the early eighteenth century, when Russia's rulers began adopting European models of culture and statecraft. The country's opening to the West involved a growing interest in foreign approaches to the governance of nature. The administration in Saint Petersburg drew heavily on the traditions and principles of German forestry when first designing forest conservation measures.[5] Like in many other parts of early modern Europe, it was the experience of the Low Countries that inspired ideas to transform and "improve" Eurasia's wetlands. Upon learning how the Dutch had expanded their farmland through drainage and how peat had fueled the country's urban and industrial growth, scientists and bureaucrats pondered the potential of wetlands in their own country.[6] Returning to Russia from his famous journey to Western Europe, emperor Peter I ordered that peat be extracted in the region of Voronezh to provide fuel for the shipyards of his nascent Black Sea fleet. He also granted several individual privileges to foster the exploration and exploitation of peat deposits in other parts of his empire.[7]

A testimony to Peter's admiration for the Dutch, these measures also reflected the tsar's growing preoccupation with forest conservation at a time when the construction of navy ships and the export of timber put substantial pressure on Russia's woodlands. The calculus was simple.

[5] Stephen Brain, *Song of the Forest: Russian Forestry and Stalinist Environmentalism, 1905–1953* (Pittsburgh: University of Pittsburgh Press, 2011), 14–17.

[6] Simona Boscani Leoni, "Die Debatte um den Torfabbau im 18. Jahrhundert: Die Gebrüder Scheuchzer zwischen Johannes von Muralt und Johann I. Bernoulli," *Jahrbuch für Geschichte des ländlichen Raums* 14 (2017): 81–95; Esa Ruuskanen, "Valuing Peatlands and Wetlands: Mires in Natural Resource and Land Use Policies in the Nordic Countries from the Late Eighteenth Century to the Present Day," in Gordon Winter and Andreas Dix, eds., *Trading Environments: Frontiers, Commercial Knowledge and Environmental Transformation, 1820–1990* (New York: Routledge, 2016), 118–136. On land reclamation and peat extraction in the Netherlands, see Jan de Vries and Ad van deer Woude, *The First Modern Economy: Success, Failure and Perseverance of the Dutch Economy, 1500–1815* (Cambridge: Cambridge University Press, 1997), 27–45; Richard W. Unger, "Energy Sources for the Dutch Golden Age: Peat, Wind, and Coal," *Research in Economic History: A Research Annual* 9 (1984): 221–253.

[7] N. Ia. Nikitinskii, *Torf i ego razrabotka na toplivo* (Moscow: Izdanie politekhnicheskogo obshchestva, 1882), 3; L. V. Kopenkina, *Istoriia torfianogo dela v Rossii* (Tver': TGTU, 2015), 16–18.

As fewer trees would be needed to produce heat when burning peat, promoting peat fuel complemented the state's simultaneous and much more ambitious attempts to sustain forests. A few decades after Peter began such efforts, polymath Mikhail Lomonosov made this reasoning explicit in his *On the Strata of the Earth*, published in 1763. Animated, like Peter, by an extended stay in Western Europe, Lomonosov described in great detail how the Dutch exploited their peat bogs and how a dense network of canals allowed them to ship the peat to regions with high fuel demand.[8] The scientist left no doubt that the Dutch example held great value for his own country. Unearthing the "subterranean economic treasure" (*podzemnoe ėkonomicheskoe sokrovishche*) of the swamps would render previously useless landscapes economically valuable and prevent the depletion of woodlands and related fuel shortages.[9] At the time, neither Peter's privileges nor Lomonosov's writings caused any noteworthy increase in peat extraction. Still, they are clear examples of how fears of forest depletion and early modern visions to improve nature by converting "wastelands" into productive landscapes coalesced, creating the notion of Russia's wetlands as spaces of untapped natural resources that, once drained and exploited, would serve Cornucopian visions of expanding material wealth.[10]

Within the Russian Empire, ideas about the possible uses of wetlands began circulating more widely once commercial activities in the countryside gained momentum in the late eighteenth century. After the nobility had been released from obligatory state service in 1762, much of Russia's economic growth originated from rural regions, where gentry families got involved in market-oriented farming and proto-industrial activities, such as running distilleries, brickworks, or weaving mills.[11]

8 M. Lomonosov, *O sloiakh zemnykh i drugie raboty po geologii. S predisloviem prof. G. G. Lemmleina* (Moscow: Gosudarstvennoe izdatel'stvo geologicheskoi literatury, 1949), 33–35.

9 Ibid., 82–84.

10 On the changing manifestations of Cornucopianism in Western economic thought, see Fredrik Albritton Jonsson and Carl Wennerlind, *Scarcity: A History from the Origins of Capitalism to the Climate Crisis* (Cambridge, MA: Harvard University Press, 2023). Important case studies on the wasteland discourse and wetlands in modern Europe include David Blackbourn, *The Conquest of Nature; Water, Landscape and the Making of Modern Germany* (London: W. W. Norton, 2006), 21–45; Vittoria Di Palma, *Wasteland: A History* (New Haven, CT: Yale University Press, 2014); Michał Pospiszyl, "The Fifth Element: The Enlightenment and the Draining of Eastern Europe," *Environmental History* 28, no. 2 (2023): 361–388.

11 Stanziani, *After Oriental Despotism*, 96–97.

Contributing to the rising economic influence of the nobility were inviolable private land titles, which, from 1782 onward, also included forests, water, and minerals located on private territory. This "enclosure of natural resources," as the historian Ekaterina Pravilova termed it, created an incentive for landowners to search for "reasonable" land uses that best suited their commercial interests.[12] The gentry's desire to extract rent from their property also extended to wetlands within their domains. While most searched for ways to expand the land suitable for farming, some also developed an interest in exploiting wetlands to generate energy. Prince Petr P. Dolgorukov, for instance, who located a peat deposit some kilometers east of the city of Moscow, experimented with burning peat in the kilns of his glass and brick factories.[13] In the early 1790s, when a landowner discovered peat on his property in the province of Smolensk, this inspired speculations about regular peat shipments to Moscow and Saint Petersburg.[14] Although peat was not at the center of the enlightened discourse about the estate economy, these early reports illustrate that the growing economic activity of Russia's landowning class inspired a way of thinking which reduced peatlands to those features that could be converted into abstract value or a means to produce tradable goods.

Even so, for almost a century, knowledge about peat's possible applications and potential market value did not boost considerable extractivist activities. Unlike the prospect of developing agriculture on reclaimed wetlands, an idea which was popular in other parts of Europe at the time and closely matched Enlightened aspirations to improve nature, extracting fuel from marshy plots of land remained unpopular. The *Transactions* (Trudy) of the Free Economic Society published detailed instructions about how to localize, extract, and burn peat as early as

[12] Ekaterina Pravilova, *A Public Empire: Property and the Quest for the Common Good in Imperial Russia* (Princeton, NJ: Princeton University Press, 2014), 24–25. On the notion of property in the Russian Empire, see also Martina Winkler, "'Mein Besitz, Landgut, Erbland, Dorf oder wie auch immer Du es nennen möchtest': Eine russische Begriffsgeschichte des Eigentums," *Jahrbücher für Geschichte Osteuropas* 60, no. 3 (2012): 321–349.

[13] N. Sokolov, "Opisanie novykh okolo Stolichnago goroda Moskvy priiskov torfa v 1794 godu," *Trudy Vol'nogo ėkonomicheskogo obshchestva* no. 3 (1798): 76–83; "Pis'mo Ego Siiatel'stvo Kniazia Petra Petrovicha Dolgorukova o torfe i kamennykh ugliakh im otkrytykh," *Trudy Vol'nogo ėkonomicheskogo obshchestva* no. 3 (1798): 68–75.

[14] F. Tumanskii, "O torfe," *Trudy Vol'nogo ėkonomicheskogo obshchestva* no. 18 (1793), 236.

1766.[15] Until well into the nineteenth century, however, few landowners considered peat a relevant resource. Given that its extraction was laborious and that burning it required special ovens and produced an intense smell, peat constituted an emergency fuel which became attractive in times of sudden price spikes in timber markets. In the late eighteenth century, for instance, the high demand for construction materials and firewood in Moscow and Saint Petersburg made contemporaries interested in exploiting the marshlands in the cities' environs to escape the bottlenecks in regional fuel markets.[16] Still, even though members of the gentry were learning that the muddy soil of Russia's swamps could produce flames hot enough to fuel their economic ambitions, enthusiasm for the fuel remained low. Expectations to profitably sell peat to urban consumers did not yield considerable success either. Two entrepreneurs, who invested some 45 thousand rubles into setting up a company to exploit peatlands close to the city of Moscow in 1839, soon quit their business because they could not find any buyers.[17] As long as timber was available and cheap enough to operate manufactures and keep houses warm, interest in peatlands as fuel deposits remained ancillary to reclaiming them for farming.

The imperial state was crucially involved in the appropriation of Eurasia's peatlands. While Peters I's idea to introduce peat as a fuel never inspired more than a few uncoordinated attempts to locate and burn peat, the central administration became more invested in exploring and exploiting peatlands for energy during the nineteenth century. Related efforts, at first, sought to address particular local problems. In the province of Moscow, where timber shortages had occurred regularly since the late eighteenth century, administrative measures to protect woodlands and the promotion of peat fuel were two sides of the same coin. In 1839, Tsar Nicholas I ordered the newly established Ministry of State Domains to prospect for alternative fuel sources close to the city to prevent further forest depletion and timber price increases. Nicholas had good reasons to

[15] Leman', "O turfe i o perezhiganii onago v ugol'e," *Trudy Vol'nogo ėkonomicheskogo obshchestva* no. 2 (1766): 29–54; "Izvestie o vygodneishei postroike pechi dlia obzhiganiia kirpichei torfom," *Trudy Vol'nogo ėkonomicheskogo obshchestva* no. 3 (1789): 86–113.

[16] On the fuel crisis and the search for "fossil alternatives" in late eighteenth-century Moscow and Saint Petersburg, see Evgeniia Lupanova, *Istoriia zakreposhcheniia prirodnogo resursa: Lesnoe khoziaistvo v Rossii 1696–1802 gg.* (Saint Petersburg: EUSP Press, 2017), 293–296.

[17] Nikitinskii, *Torf i ego razrabotka*, 6–7.

be concerned. After Moscow had been set on fire and abandoned during Napoleon's advance in 1812, reconstructing the city required large amounts of wood. Manufacturing businesses, particularly textile industries, added further pressure on the region's forests. Following the Tsar's order, a permanent turf master (*torfmeister*) was assigned to oversee the exploration and appropriation of state-owned wetlands in the province of Moscow.[18]

The creation of the post reflected the imperial administration's changing role in the governance of Russia's natural wealth. During Nicholas' reign, the Ministry of State Domains began to carve out and enforce power over the state's property. Forests, the most important sources of fuel and construction materials, were at the center of the administration's growing interest in natural resources.[19] Industrial policies now began to mirror anxieties about threatened forests and the idea that using peat as industrial fuel could be a remedy. In the early 1840s, the Ministry of State Domains and the Ministry of Finance debated the future of distilleries in the province of Moscow. The senate had prohibited the creation of new distilleries in 1796 in response to Moscow's skyrocketing fuel prices. Half a century later, when the timber market remained strained, the Committee of Ministers suggested that the restrictions could be lifted if new distillers would be obliged to base liquor production solely on peat fuel. Any violation of this rule would lead to the closure of a distillery and the confiscation of its property by the treasury.[20] Prescriptive resource-use policies affected other industries as well. During the 1850s, the Moscow military general governor headed a newly created special committee which pursued peat extraction and attempted to teach industrial producers the benefits of peat.[21]

As manufacturers remained suspicious about the fuel, administrative pressure at times complemented the monetary incentives and practical advice offered to convince them to switch to peat. Ivan I. Vikhliaev, a leading expert on peat matters in the late imperial administration, commented later that peat at that time "had to be sold almost forcibly" to industrial plants. In some cases, factory owners even bought the

[18] "Ob otvrashchenii istrebleniia lesov upotrebleniem torfa vmesto drov," *Zhurnal Ministerstva Gosudarstvennykh Imushchestv* chast' 1 (1841): 109–134.

[19] Pravilova, *A Public Empire*, 47–58.

[20] "Otnositel'no ustroistva vinokurennykh zavodov v Moskovskoi gubernii" (1844), RGIA f. 1263, op. 1, d. 1644, ll. 87-93. I thank Igor' Kuznetsov for drawing my attention to this case.

[21] Nikitinskii, *Torf i ego razrabotka*, 9.

committee's peat just to throw it away, keeping some bits in the furnace rooms to present them in case of an inspection.[22] Another contemporary pointed out that state institutions at times employed "compulsory measures" (*prinuditel'nye mery*) to foster peat extraction and that industrial producers initially adopted peat fuel against their will.[23]

The state's sometimes forceful attempts to stimulate peat extraction were embedded within a broader discourse about the value and meaning of nature. From the 1840s onward, the "Forest Question" occupied the minds of government officials, landowners, scientists, artists, and writers who feared that excessive logging would threaten the survival of Russia's forests. At this juncture, the idea that woodlands needed human care had had a long history in the empire.[24] In the nineteenth century, however, concerns about deforestation carried a new sense of urgency. Linking anxieties about resource depletion with questions of Russian spiritual and national identity, the manifold publications surrounding the Forest Question promoted forest conservation as an economic necessity and a moral responsibility.[25] Against this backdrop, the exploitation of peatlands appeared as an indirect instrument for conserving economically valuable resources and, simultaneously, as a means to retain the essence of the Russian landscape.[26] Specialist publications, such as the *Journal of the Ministry of State Domains* and the *Forestry Journal*, regularly drew a link between peat extraction and the conservation of forests.[27] Mikhail Astrov, the physician and outspoken advocate of forest conservation in Anton Chekhov's famous play *Uncle Vanya*, also made this connection. When told that people needed wood to build and heat their houses, Astrov replies: "You can burn peat in your stoves and build your sheds of stone" before listing all the damages that excessive use of wood can

[22] Vikhliaev, "Kratkii ocherk razvitiia kul'tury bolot i utilizatsii torfa v otdel'nykh gosudarstvakh," *Vestnik Torfianogo Dela* 1, no. 3 (1914), 255–256. See as well Kopenkina, *Istoriia torfianogo dela*, 54.

[23] Nikitinskii, *Torf i ego razrabotka*, 10.

[24] For a detailed account, see Lupanova, *Istoriia zakreposhcheniia prirodnogo resursa.*

[25] Jane Costlow, *Heart-Pine Russia: Walking and Writing the Nineteenth Century Forest* (Ithaca: Cornell University Press, 2013), 81–114.

[26] A similar discourse evolved in Mexico, where the use first of coal and later of oil was framed as a means to protect the country's forests from depletion. See Germán Vergara, *Fueling Mexico: Energy and Environment, 1850–1950* (Cambridge: Cambridge University Press, 2021), 90–93, 157.

[27] See, for instance, "O turfe," *Lesnoi zhurnal* 4, no. 2 (1833): 117–129; "Razrabotka torfa po vedomstvu lesnogo departamenta," *Zhurnal Ministerstva Gosudarstvennykh Imushchestv* 14, no. 2 (1854): 97–100.

cause.[28] Astrov's words are typical of how nineteenth-century imperial elites saw wetlands. Promoting the idea that extracting peat represented a way around the resource dilemma facing central Russia's provinces, they constructed a hierarchy of ecosystems which would inform the cultural perceptions and practical uses of peatlands for at least a century to come: Peatlands should be exploited for the sake of forests.

BUREAUCRATIC VISIONS OF PROFIT AND DEVELOPMENT

The state's position as a landowner profoundly shaped the appropriation of peatlands, particularly once the central administration assumed an increasingly active role in governing the empire's natural wealth. The Ministry of State Domains, created in 1837, embodied the state's changing approach to the nonhuman world. Linking cameralist understandings of profitability (*vygoda*) with the aim of responsible resource use, the ministry put extensive efforts into surveying state domains. It also rented out state property or sold licenses for temporary exploitation. Issuing *obrochnye stat'i*, as these titles were officially called, allowed the Ministry of State Domains and its institutional heirs to extract fiscal revenue from state property, be it farmland, forests, or fisheries, while at the same time retaining control over how intensely it was used.[29] The imperial land administration also played an active role in commodity markets. At the end of the nineteenth century, the ministry's Forestry Department gained considerable income from selling timber to domestic businesses and foreign customers.[30]

Wetlands constituted a considerable part of the state's land holdings. As the imperial administration became more invested in turning this property into a source of income, these lands attracted the attention of bureaucratic elites who subscribed to the trope of wetlands being underused "wastelands." The creation of a Turf Master's Division (*Torfmeisterskaia Chast'*) within the Ministry of State Domains in 1883 turned peatlands into regular objects of state governance. Appointed to

[28] "Uncle Vanya," in Anton Chekhov, *Plays*. Translated with Notes by Peter Carson. With an Introduction by Richard Gilman (London: Penguin Books, 2002), 154.

[29] On the transformation of fisheries into *obrochnye stat'i* in the Aral Sea region, see Niccolò Pianciola, "Cossacks and Sturgeons: Fisheries, Colonization and Science around the Aral Sea (1873–1906)," *Journal of the Economic and Social History of the Orient* 62, no. 4 (2019): 626–673.

[30] Kiril Tomoff, "The Role of Forests in Witte's Industrialization Drive," *Russian History* 22, no. 3 (1995): 249–283; Brain, *Song of the Forest*, 22.

run this new unit was Leonid A. Sytin (1829–1913), a graduate of the Agricultural Institute in Gory, who had previously served as the Ministry's first turf master and would maintain his role as the head of the Turf Master's Division until he died in 1913.[31] During his tenure, the unit grew from an initial staff of 6 members to 24 in 1914. Its annual budget increased steadily as well. In the early twentieth century, the developments within the central administration also reverberated in the provinces, as regional branches of the Ministry (since 1905, known as the State Administration of Land Settlement and Agriculture, GUZiZ) began appointing turf masters. At the beginning of the First World War, the provinces of Saint Petersburg, Moscow, Vladimir, Nizhny Novgorod, Mogilev, and Tambov all had their own turf master who monitored and coordinated how wetlands were being put to use.[32]

Once they were identified as a distinct unit of land governance, wetlands became subject to the same principles of resource management that the imperial administration deployed to other parts of nature in state ownership as well. Preparing peatlands for rental lease and declaring them *obrochnye stat'i* were the key tasks which Sytin and his staff carried out during their day-to-day. That this required the application of the homogenizing language of money to ecologically diverse wetland environments does not seem to have troubled them. In their role as administrators of state property, the officials in the Turf Master's Division acted very much like the landed gentry, seeking to yield the highest possible return from their holdings. The notions of profitability (*vygoda, vygodno*) permeating the unit's records drew upon a longer tradition of valuing and monetizing the nonhuman word. Indeed, members of the nobility had used modern concepts of profit in relation to their privately owned forests since the late eighteenth century.[33] Although leases of peatlands or tickets for their temporary exploitation never added significantly to the state budget, they secured a small but steady flow of income for the imperial

[31] "L. A. Sytin," *Vestnik Torfianogo Dela* 1, no. 1 (1914): 9–10.

[32] GUZiZ, *Otchet o rabotakh Torfmeisterskoi chasti v 1913 god* (Petrograd: Tipografiia Al'tshulera, 1915); I. I. Vikhliaev, "Kratkii ocherk," 258.

[33] Elena Korchmina, "Conceptualising the Notion of 'Profit' among the Russian Nobility (Second Half of the 18th–First Half of the 19th Centuries)," *Quaestio Rossica* 6 no. 4 (2018): 1144–1159. For examples of this rhetoric in the records of the Turf Masters's Division, see for instance the Letter from the Head of the Turf Master's Division to the Department of Land Improvement (1889), RGIA f. 425, op. 1, d. 2, l. 67; Letter from the Land Administration of Moscow and Tver' to the Peat Division (1909), RGIA f. 425, op. 1. d. 35, l. 1; Letter from the Land Administration Tambov (1912), RGIA f. 425, d. 38, l. 8.

administration.[34] Most payments came from the central Russian provinces of Vladimir, Moscow, and Tver. The Morozov cotton and weaving factories at the border of the Moscow and Vladimir regions exploited state-owned wetlands continuously from the early 1880s onward. Other industrial producers followed along the same route. In 1914, private users utilized a little over 100 state-owned peatlands for lease periods of sometimes up to 60 years.[35]

But how much was a peatland worth? Putting a price tag on a specific piece of wet land was by no means a straightforward and, in fact, quite an arbitrary process. When imperial officials created *obrochnye stat'i*, they usually took the estimated market value of the yield from a state-owned asset to define payment rates for its use: the price of wood in the case of a forest, of grain in the case of a field, or of fish in the case of fishing grounds. Peat, on the other hand, was not a traded commodity when the Turf Master's Division began to rent out the state's wetlands. Making these environments available for exploitation therefore required a bureaucratic procedure called "primary valuation" (*pervonachal'naia otsenka*), during which state officials sought to balance fiscal interest and estimated consumer demand with the peculiar characteristics of a given deposit and the quality of the peat it contained. In 1884, for instance, Sytin and his staff singled out the First Moss Peatland (*1-e Mokhovoe boloto*) northwest of Moscow as a promising site for extraction. Having praised the high quality of the peat and the wetland's proximity to a silk and carpet factory, a cotton mill, and a distillery, their report reflected on the timber price in the region to finally comment on the varying quality of the peat. Eventually they established a lease rate of 25 kopeks per cubic sazhen of dry peat. Things looked similar in another deposit in the area. Located 10 versts from the railways but without potential buyers nearby, the Second Moss Peatland was valued at 20 kopeks per kubik sazhen of peat. In contrast, the price for the same

[34] In 1913, *obrochnye stat'i* issued by the Turf Master's Division earned the fiscus some meagre 124,415 rubles, less than 1 percent of the income generated from all *obrochnye stat'i*. The overall revenue from *obrochnye stat'i* in 1913 was 20,1 mio rubels. See GUZiZ, *Obzor deiatel'nosti za 1913 god* (Petrograd: Tipografiia Sel'skogo Vestnika, 1915), XV.

[35] "Spisok torfianykh bolot sdannykh v soderzhanie po Vladimirskoi gubernii," RGIA f. 425, op. 1, d. 2, ll. 103–105; "O bumagakh ne trebuiushchikh dal'neishago proizvodstva (1909–1914)," RGIA f. 425, op. 1, d. 35, ll. 104; "O rabotakh Torfmeisterskoi chasti v 1914–1916 gg.," RGIA, f. 425, op. 1, d. 43, ll. 207–240; Vikhliaev, "Kratkii ocherk," 266.

amount of first-class peat from the Svitinskii Peatland in the Tver province ranked at only 10 kopeks because the deposit was surrounded by private land, hard to access, and quite far from the railways.[36] "Primary valuation," reports like this demonstrate, represented a choice to define a wetland in relational rather than material terms, making complex wetland ecosystems subject to the simplifying gaze of monetary valuation.

Lease rates for state-owned wetlands varied not only across places but also across time. Acquiring permission to exploit state-owned peatlands in the province of Moscow became increasingly costly in the late imperial period. In the district of Bogorodsk, such payments rose from 34,90 kopeks per cubic sazhen in 1892 to 73 kopeks in 1913. Rental charges in the district of Klin northwest of the city of Moscow more than tripled between 1894 and 1908.[37] Changing lease rates did not reflect the motions of an "invisible hand" but the decisions of bureaucrats who tried to extract as much fiscal revenue from a given piece of land as possible. In 1915, the administration of the Balashov Spinning and Weaving Mills in Ivanovo approached the forestry administration of the Moscow and Tver region to enquire whether they could rent an unexploited state-owned peat deposit. Though principally agreeing with the request, the official responding to the enquiry emphasized that rent levels must reflect recent price trends. As peat prices had been showing considerable increases, the rental charge should be at least as high as what the company was already paying for the permission to exploit another state-owned wetland.[38] Reflections like these underscore the blurred boundaries between state and market at a time when economic change challenged the foundation of Russia's social order while the meaning of state property was in flux. The imperial administration acted as both a creator of and a participant in the market for peat. Officials not only converted the labor of the earth into abstract monetary value. Acting on behalf of the state's fiscal interest, they also actively exploited the opportunities arising from the growing commercial demand for fuel and the increasing appreciation for peatlands as energy deposits.

Raising income for the state was not the only aim the Turf Master's Division sought to achieve. Moving beyond narrow fiscal aims, the state's

36 "Otchet o rabotakh po issledovaniiu bolot za 1884 god," RGIA f. 425, op. 1, d. 3, ll. 34–53.

37 *Trudy Kommissii po izucheniiu torfianogo dela* (Moscow: Obshchestvo dlia sodeistviia uluchsheniiu i razvitiiu manufakturnoi promyshlennosti, 1916), 84.

38 "O rabotakh Torfmeisterskoi chasti v 1914–1916 gg.," RGIA f. 425, op. 1, d. l. 20.

engagement with peatlands also reflected a broader commitment to support industrial expansion. Most lease titles for state property were designed to reconcile competing claims and prevent resource depletion. In the case of peatlands, however, they served as an injection to stimulate resource use. Parties interested in exploiting state-owned assets typically had to make a bid in a public auction, after which the highest bidder would be granted permission to move ahead. From 1884, in line with the state's increasing commitment to support peasant farming following the abolition of serfdom in 1861, the law made an exception for peasant communities, allowing them to lease state land without competitive bidding (*bez torgov*). In 1896, the same approach was extended to peasants interested in purchasing logging tickets for state forests.[39] When it suited the broader aim of promoting industrial growth, the Turf Master's Division sometimes diverted from hosting public auctions, offering companies interested in extracting peat from state-owned land leases at a preset rate (*bez torgov*). The Nikol'skaia cotton mill, for instance, exploited wetlands in the west of the Vladimir province under such terms. In one case, a private entrepreneur whose company experimented with the carbonization of peat got to exploit a state-owned wetland without an auction.[40] For the administration, these leases were a way to combine profitable land management with measures to provide industrial producers with a cheap energy source. With regards to peatlands, state and business pursued increasingly overlapping interests.

The idea that the imperial administration should promote the profitable exploitation of wetlands became more pronounced after the turn of the century, as members of the Turf Master's Division increasingly worked on landholdings which did not belong to the state. "To accommodate consumers' needs and to extract the highest possible benefit (*vygoda*) for the treasury through the exploitation of peat deposits," as an official document stated in 1909, state-employed engineers and technicians regularly surveyed land upon private request.[41] While the turf master of the province of Moscow had already received permission to do so in the 1850s, such activities were of secondary importance as long as

[39] S. I. Ilovaiskii, *Uchebnik finansovogo prava. 4-e izdanie* (Odessa: Tipo-Khromo-Lit. A. F. Sokolovskago, 1904), 86–88; Brain, *Song of the Forest*, 49–50.

[40] "Spisok torfianykh bolot sdannykh v soderzhanie po Vladimirskoi gubernii," RGIA f. 425, op. 1, d. 2, ll. 103–105.

[41] Letter from the land administration in Moscow and Tver' (1909), RGIA f. 425, op. 1, d. 35, l. 1.

industrial demand for peat remained low.[42] This changed in the early twentieth century when manufacturers and railway managers tried to reduce their reliance on ever more expensive timber by using other fuel sources.[43] Once landowners became aware that unearthing the dormant energy stored in their wetlands could provide them with a source of income, they began to appeal to the Department of Agriculture for assistance. Take the nobleman Viacheslav Bulychev from Kostroma, who was hopeful that industrial demand for peat would allow him to raise some income from the wetlands on his estate: "Wishing to utilize (*utilizirovat'*) these swamps, I dutifully request that a turf master be sent to determine the properties of these swamps," he wrote to the State Administration of Land Settlement and Agriculture in May 1914.[44] Ivan Vikhliaev, who had just succeeded Sytin as the head of the Turf Master's Division, noted in the same year that he and his staff could barely keep up with the high number of private requests for wetland surveys that usually preceded formal lease or sales agreements.[45]

Turning peatlands into extraction sites was even believed to present a service to the broader common good. In 1914, the Department of Agriculture adopted new rules on the secondment of technicians upon request. State experts charged with the valuation of a peatland had to conduct a general survey of the land and its surroundings, take soil samples, and assess the quality and volume of the peat. The new rules detailed how the administration intended to charge landowners for such services. Whoever wished that their land be surveyed would have to cover daily allowances for the ministry's technician and related travel costs. The instructions made some significant exemptions though. If peasant communes and rural self-governing organs known as zemstvos wished to promote peat extraction "for common goals" (*obshchepoleznye tseli*), they would have to take care of any locally arising labor or material needs. They would not, however, have to reimburse the state-employed technician.[46]

42 "O razreshenii torfmeisteram zanimat'sia v chastnykh imeniiakh," *Zhurnal Ministerstva Gosudarstvennykh Imushchestv* 18, no. 1 (1858): 58–59.

43 E.g. Letter from the Department of Agriculture to the head of the Peat Division (1884), RGIA f. 425, op. 1, d. 3, ll. 1–2; Note on a letter from the Riazan'–Ural Railways and their interest in peat along the railway (1900), RGIA f. 425, op. 1, d. 2, l. 124.

44 Bulychev to the Chief Administration of Land Settlement and Agriculture (1914), RGIA f. 425, op. 1, d. 43, l. 2.

45 Vikhliaev, "Kratkii ocherk," 270–271.

46 "Pravila komandirovaniia chinov Torfmeisterskoi chasti glavnogo upravleniia zemleustroistva i zemledeliia po khodataistvam chastnykh lits" (1914), RGIA f. 425, op. 1, d. 4, ll. 13–19.

The document did not specify what constituted such "common goals," but it was still a firm indicator of the various aims embedded in the administration's approach to governing Russia's natural wealth. While the central bureaucracy acted like a private landowner interested in monetary gain on the one hand, it increasingly endorsed the extension of public services, including land assessment, as a public duty on the other.

COMING TO TERMS WITH ABUNDANCE

Its abundance was one of the main reasons making peat so promising. From the late eighteenth century, when peat became a subject of the Enlightenment discourse, until the twentieth century, the seemingly ubiquitous nature of wetlands was one of the most influential tropes shaping elite perceptions of the central Russian landscape. Marshy and impassable lands had no place in the Arcadian and picturesque landscape aesthetics that dominated Russian culture in the late eighteenth and early nineteenth centuries. Moreover, unlike forests, which became a marker of Russia's national landscape during the nineteenth century, peatlands remained part of the "meager nature" that the empire's elites struggled to endorse as a source of inspiration and identity.[47] Recurring worries about the deteriorating state of central Russia's forests were most conducive to the idea that peat existed in unlimited quantity and was bound to become a vehicle of economic development. In 1798, Prince Dolgorukov noted that peat existed in abundance (*preizobil'no*) in the region of Moscow, indicating its potential to become a regular source of heat.[48] Imaginaries of plenty remained vital in the wetlands discourse well into the twentieth century. In 1914, Ivan Vikhliaev, head of the Turf Master's Division, projected that peat could sustain Russia's long-term need for fuel: "If all combustibles on the market were replaced by peat, it would last for a couple of centuries."[49]

For a long time, however, the abundance of peat was a hypothesis rather than a proven fact. No more than a few dots along the Gravorna, nowadays one of the many underground rivers in the city of Moscow, mark peat deposits on a map which the *Transactions* of the Free Economic Society published shortly after Dolgorukov made his peat

[47] Christopher Ely, *This Meager Nature: Landscape and National Identity in Imperial Russia* (DeKalb: Northern Illinois University Press, 2002).

[48] "Pis'mo Dolgorukova," 72.

[49] Vikhliaev, "Kratkii ocherk," 263.

discovery in the early 1790s (Figure 1.1).[50] When the journal of the Ministry of State Domains announced in 1841 that a permanent turf master would take up his work in the province of Moscow, it provided a map which indicated locations suitable for extraction. The map identified a cluster of relevant peat deposits near the town of Bogorodsk, where factory owners would set up extensive peat works later in the century and where Russia's first peat-based power plant *Ėlektroperedacha* started producing electricity shortly before the First World War. However, even though peat soils are prevalent in many parts of the province of Moscow, the cartographer does not seem to have known their exact size or location. Most parts of the territory appear as terra incognita on the map. The practical interest which motivated its creation in the first place was undeniable though. The map's legend distinguished peat deposits based on how difficult their drainage and exploitation would be, thus reducing peatlands to resources instead of appreciating them as elements of a complex and interconnected natural world (Figure 1.2).

Not all peat, moreover, was the same. Peat is a heterogeneous matter and the product of the specific conditions under which a peatland formed over time: the composition of plants that had slowly decayed into peat, the soil on which these plants had been growing, the deposition of inorganic substances from the atmosphere in the ground.[51] Those interested in exploiting peatlands for fuel soon realized that peats of different sorts had different capacities to burn and produce heat. Before any extraction could start, they had to establish whether or not the peat in a given place would hold up to the expectations raised by scholars, such as Lomonosov. Tracing the historical attempts to assess and classify peat not only unearths the gradual evolution of scientific knowledge about peat soils in imperial Russia. It also reveals an associated process of abstraction by which contemporaries began to value peat separately from its living environment and described it in homogenizing terms which reflected human needs rather than the ecological complexity of peatlands.[52]

When the chemist Lovits received two samples of Prince Dolgorukov's peat for examination in 1793, the methods for such studies were still in flux. In his detailed report, Lovits described the flame and smoke of

50 Sokolov, "Opisanie novykh okolo Stolichnago goroda Moskvy priiskov torfa," 76.

51 R. Kelman Wieder, Merritt R. Turetsky, and Melanie A. Vile, "Peat as an Archive of Atmospheric, Climatic and Environmental Conditions," in Edward Maltby and Tom Barker, eds., *The Wetlands Handbook* (Chichester: Wiley-Blackwell, 2009), 96–112.

52 Noel Castree, "Commodifying what Nature?," *Progress in Human Geography* 27, no. 3 (2003): 273–297.

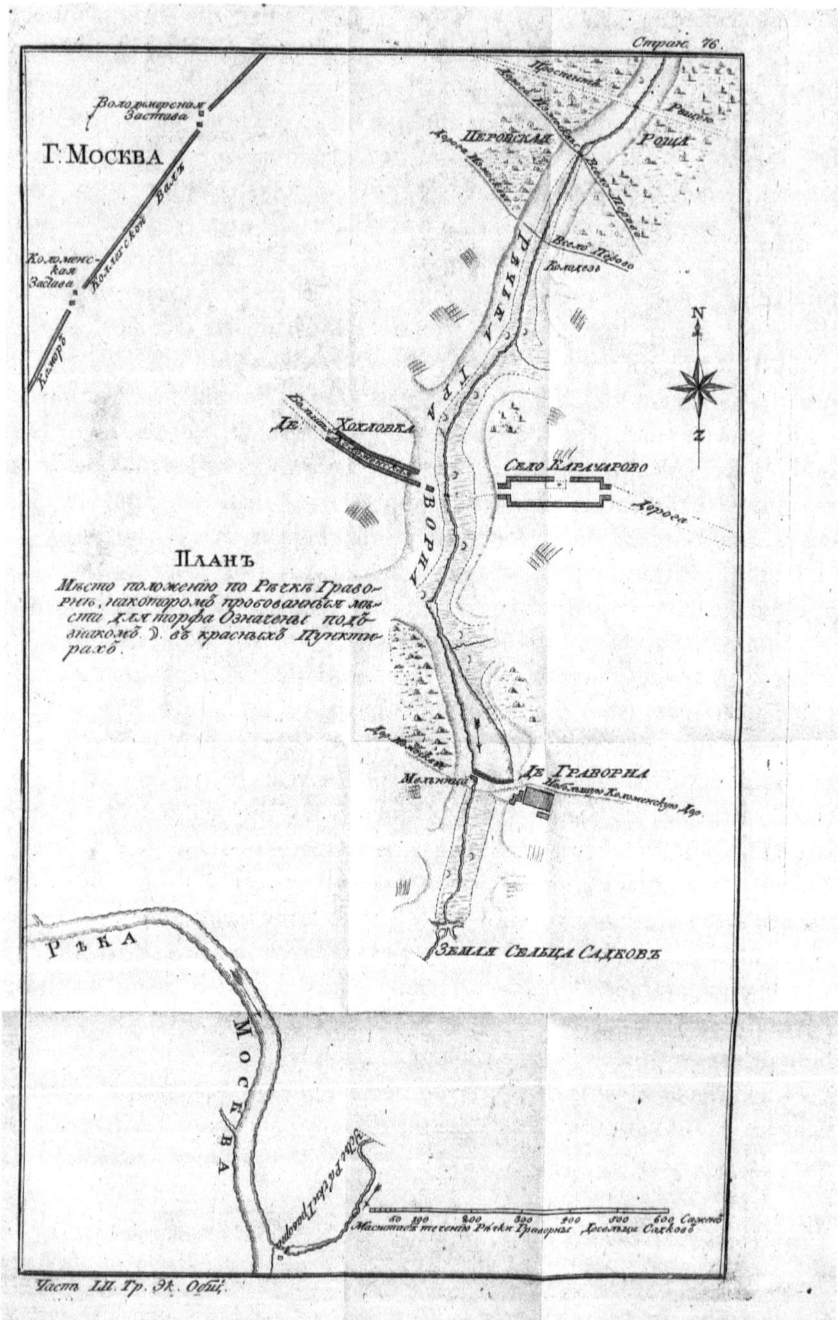

FIGURE 1.1 Peat deposits along the Gravorna River east of Moscow in the late eighteenth century. *Source*: N. Sokolov, "Opisanie novykh okolo Stolichnago goroda Moskvy priiskov torfa v 1794 godu," *Trudy Vol'nogo ėkonomicheskogo obshchestva* no. 3 (1798).

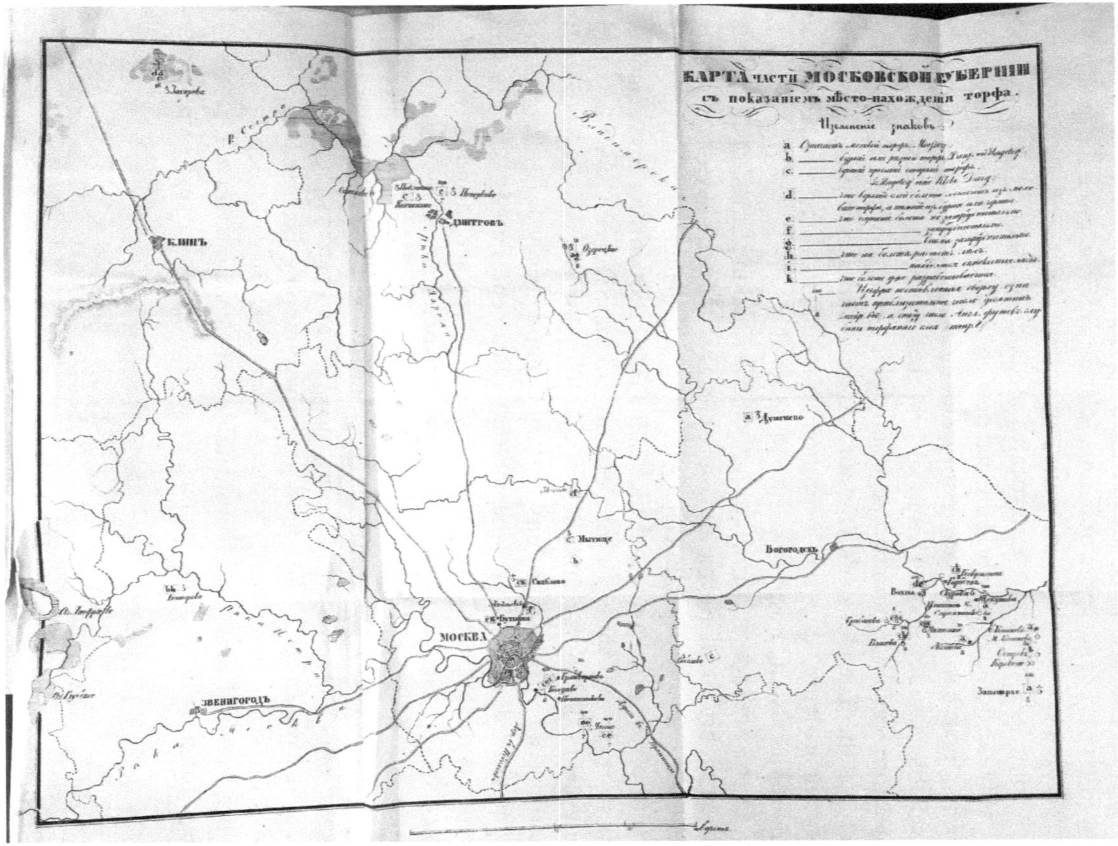

FIGURE 1.2 Peat deposits in the Province of Moscow in the first half of the nineteenth century. *Source*: "Ob otvrashchenii istrebleniia lesov upotrebleniem torfa vmesto drov," *Zhurnal Ministerstva Gosudarstvennykh Imushchestv* chast' 1 (1841).

burning peat, using these criteria to establish how suitable the samples were for commercial use.[53] Over time, such observations lost their significance. In the early twentieth century, modern chemistry had largely replaced experience-based evidence of the kind Lovits had produced with the help of his hands, eyes, and nose. Scientists now assessed the quality of a specific peat sample by measuring its carbon, hydrogen, nitrogen, and ash content.[54] But these sophisticated methods to analyze, describe, and classify the nonhuman world created new challenges, as fine-tuned instruments, most importantly scales for exact measurements, were not widely available. Landowners or regional branches of the land administration therefore regularly submitted peat samples for examination in the laboratory of the Turf Master's Division. As peatlands became an object of economic desire, imperial institutions turned into a space where local environments and abstract science intersected.

In the absence of airplanes and satellites, taking stock of peatlands required, first of all, people and time. The Turf Master's Division typically sent out one or two technicians who surveyed the terrain, estimated the depth of the peat layer, and took soil samples for further investigation. As a rule, they hired local peasants for assistance. Once they had inspected a plot and established its suitability for extraction, the officials portioned it into smaller units "to make renting easier for entrepreneurs with limited means," as one internal document stated.[55] In the late imperial period, a technician could survey some six to nine desiatinas per day, a trivial area considering the territory covered by peat soils in European Russia.[56] Marrying limited capacities and pragmatic interests, the Turf Master's Division focused on territories close to railway lines, which were more accessible for the surveyors and better suited to ship the peat to consumers.[57] The most intensive mapping occurred along the Moscow–Saint Petersburg, Moscow–Smolensk, and Moscow–Nizhny Novgorod railways. Still, considerable gaps remained in the peatland inventories even in those regions, as no systematic surveys of privately

[53] "Donesenie o Moskovskom turfe," *Trudy Vol'nogo ėkonomicheskogo obshchestva* no. 52 (1798): 113–115.
[54] *Trudy Kommissii po izucheniiu torfianogo dela*, 7–27.
[55] Report on the investigation of peat deposits on state-owned estates (1884), RGIA f. 425, op. 1, d. 3, l. 32.
[56] "Torfodobyvanie, soderzhanie chinov Torfmeisterskoi chasti i drugie raskhody po torfiannomu delu" (ca. 1914), RGIA f. 425, op. 1, d. 43, ll. 193–196.
[57] "O naznachenii rabot na 1883 god," RGIA f. 425, op. 1, d. 1, ll. 1–2.

owned or peasant lands were carried out. When a certain technician named Fedor Postnikov approached the Department of Agriculture in 1915 asking for a list of all peatlands along the Moscow–Saint Petersburg Railway, the central administration could not help him. The Turf Master's Division lacked information about privately owned peatlands, their owners, or their distance from railway stations, Postnikov learnt.[58] In the same year, Ivan Vikhliaev lamented that systematic research concerning peat "as a product of natural history and an object of technical use" remained underdeveloped.[59]

The efforts to take stock of peatlands reflected a broader environmental turn in late imperial map-making. During the nineteenth century, when scientists developed an interest in the interplay between soil, climate, and vegetation, cartographic depictions of the empire increasingly stressed environmental factors. Prominent examples include the works by the soil scientist Vasilii Dokuchaev or the topographical depictions of Siberia focusing on what would be called *permafrost* in the twentieth century.[60] Peatlands entered the empire's geographical imagination mainly due to the efforts of the Turf Master's Division. In 1914, Vikhliaev published the first map of peatlands in what imperial elites called "European Russia," squaring the challenges presented by the vast size of the territory with the ambition to translate as much knowledge about nature as possible into the language of cartography. Instead of engaging with the messy details on the ground, Vikhliaev's map erased the remarkable variety of peatland ecosystems, instead indicating the share of peatlands in the overall territory of any given province. Based on a combination of survey information and estimates, the map revealed the state's territorial attitude toward the governance of the empire's natural wealth and the crucial role of the administrative unit of the province within this larger effort. Visually, however, the map transcended a mere territorial approach. In line with cartographers' growing recognition of environmental factors, the hachures made provinces with similar shares of peat soils appear as belonging to the

[58] Vikhliaev to Postnikov (1915), RGIA f. 425, op, 1, d. 48, l. 226.

[59] Internal note from the head of the Turf Master's Division I. I. Vikhliaev to the Department for Land Improvement (1915), RGIA, f. 425, op. 1, d. 43, l. 282.

[60] Pey-Yi Chu, *The Life of Permafrost: A History of Frozen Earth in Russian and Soviet Science* (Toronto: University of Toronto Press, 2021), 39–45; Evtuhov, *Portrait of a Russian Province*, 170–172; David Moon, *The Plough that Broke the Steppes: Agriculture and Environment on Russia's Grasslands, 1700–1914* (Oxford: Oxford University Press, 2013), 76–86.

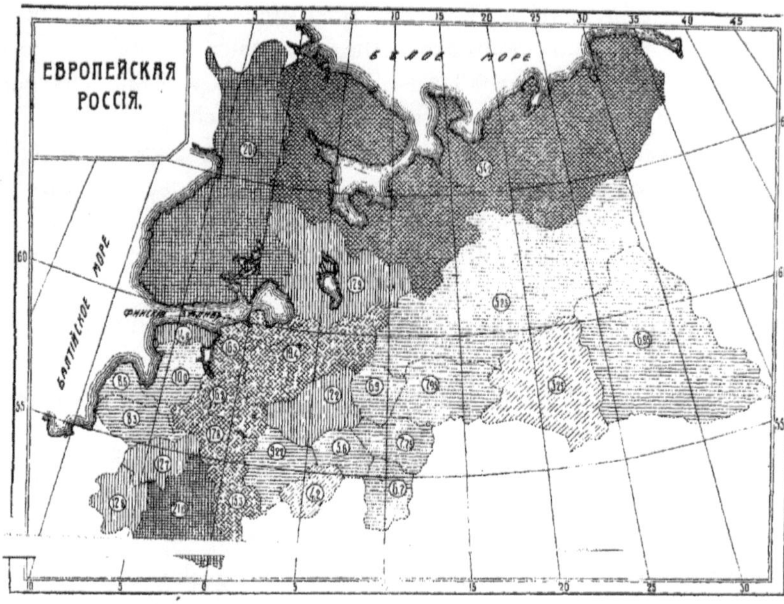

Процентное отношеніе площади болотъ къ общей площади губерніи:

- отъ 0,1% до 5%
- „ 5,1% „ 10%
- „ 10,1% „ 15%
- „ 15,1% „ 20%
- „ 20,1% „ 25%
- „ 25,1% и болѣе.

ГУБЕРНІИ.	Площадь губерніи въ десятинахъ.	Количество десят. болотъ.	Болотъ въ % къ общ. площ.
1. Московская	2.954.606	166.184	5,6
2. Тверская	5.613.041	677.484	12,2
3. Владимирская	4.209.531	327.485	7,75
4. Рязанская	3.737.353	252.165	6,7
5. Костромская	7.491.948	596.053	7,95
6. Ярославская	3.071.070	212.019	6 9
7. Псковская	3.696.461	615.715	16,8
8. Новгородская	10.087.188	1.944.244	19,4
9. Петроградская	3.787.365	624.884	16,5
10. Эстляндская	1.645.041	244.400	14,8
11. Курляндская	2.427.635	207.554	8,5
12. Минская	7.712.592	1.690.366	21,79
13. Могилевская	4.189.322	655.615	15,3
14. Гродненская	3.297.815	423 614	12,8
15 Виленская	3.525.070	449.802	12,7
16. Ковенская	3.607.420	299.484	8,3
17. Витебская	3.815.404	679.626	17,8
18. Смоленская	4.824.325	446.175	9,22
19. Калужская	2.781.372	117.249	4,2
20. Олонецкая	12.085.853	1.464.564	12,9
21. Вятская	13.963.524	450.582	3,25
22. Пермская	29.815.548	2.077.269	6,95
23. Вологодская	36.499.918	1.910.788	5,25
24. Архангельская	71 878.226	38.948.395	54,2
	246.667.128	55.481.746 ¹)	21,6
25. Лифляндская	—	—	10
26. Финляндія	—	—	20

¹) Если исключить неудобныя земли Архангельской губ., состоящія главнымъ образомъ изъ тундръ, то площадь болотъ въ 24 губ. будетъ равна 16.533.351 лесят. или 9% общей площа ти.

FIGURE 1.3 Vikhliaev's 1914 map indicating peatlands' share in the territory of the provinces in European Russia. *Source*: I. I. Vikhliaev, "Istoriia torfianogo dela v Rossii." Otd. Ottisk iz zhurnala, *Vestnik Torfianogo Dela* 1914g., no 3.

same natural space (Figure 1.3).[61] Arguably the most advanced example of wetland cartography at the time, Vikhliaev's map reproduced the

[61] On the inclusion of environmental factors in economic maps, see Nailya Tagirova, "Mapping the Empire's Economic Regions from the Nineteenth to the Early Twentieth Century," in Jane Burbank, Mark von Hagen and Anatolyi Remnev, eds., *Russian Empire: Space, People, Power, 1700–1930* (Bloomington, IN: Indiana University Press, 2007), 125–138.

notion of abundance, which had underpinned the mental and physical appropriation of Russian peatlands since the eighteenth century. In 1914, the map suggested, abundance was a proven fact. Peatlands could be found all across European Russia – and the state knew exactly how they were distributed.

THE LIMITS TO APPROPRIATION

What was the role of the peasantry within this larger resource-making process? In contrast to the advocates of peat extraction within the imperial administration and among industrial elites, most peasants valued wetlands primarily for what they offered on the surface. They used them to pick berries or mushrooms, hunt for wild birds or game, graze their livestock, or make hay. Peasants sometimes embraced the notion of "land improvement" (*melioratsiia*) to enhance their farmland or pasture. Few, however, saw peatlands primarily as potential extraction sites. For the overwhelming majority of the rural population, firewood was the most important fuel source until well into the Soviet period. As long as wood was within reach or affordable, peasants did not burn peat, which was difficult to extract and needed time-consuming drying before it could be used as a fuel.

This persisting reliance on firewood did not mean that the rising interest in peatlands as stores of fuel bypassed rural communities. State officials and modernizing elites deliberately tried to instill an interest in peat in the rural population. Their efforts were propelled by the peasantry's changing position in the social fabric of the Russian Empire following the end of serfdom in 1861, which had granted peasants personal freedom and consolidated communal land rights, though still restricting their mobility. Despite its conservative outlook, the emancipation reform had a lasting effect on the rural order and the broader imperial economy. Agricultural productivity, industrial growth, and peasant consumption developed dynamically in the late nineteenth century as peasants adapted flexibly to market and climate conditions or found employment in mines, metallurgy, industrial companies, and rural businesses.[62] Meanwhile, the central state developed a growing presence in rural affairs. In 1894, the government rebranded the former Ministry

[62] Andrei Markevich and Ekaterina Zhuravskaya, "The Economic Effects of the Abolition of Serfdom: Evidence from the Russian Empire," *The American Economic Review* 108, no. 4–5 (2018): 1074–1117; David Moon, *The Russian Peasantry: The World the Peasants Made, 1600–1930* (London: Routledge, 1999), 108–117.

of State Domains as the Ministry of Agriculture and State Domains. The reform signaled that land administration was no longer confined to managing the state's landholdings but would now entail interventionist measures to boost the rural economy. These included changes in taxation, the propagation of "scientific agriculture" through an expanding network of educational institutions and experts, and the famous 1906 Stolyin reform, which sought to weaken communal land tenure and create a class of "progressive" farmers endowed with private land titles.[63]

Given that peatlands constituted a substantial part of peasant landholdings in European Russia, peasant uses of these environments became subject to the scrutiny of state representatives and experts committed to solving the "agrarian question." In a time when many believed that "backward" peasants had to be taught how to exploit the land best and become responsible citizens, it seemed only logical to demonstrate to them that the value of wetlands derived, at least in part, from what they held underneath their surface.[64] Landowners had occasionally tried to make peasants aware of the benefits of peat fuel earlier in the nineteenth century. Yet, it was only in the context of the rising state and public interest in the countryside following the end of serfdom that peasants became targets of an information campaign that openly propagated the advantages of extracting peat. Over time, the activities of the Turf Master's Division extended considerably beyond the profitable management of state-owned wetlands or the valuation of private peatlands. Sytin, Vikhliaev, and other experts organized exhibitions and lectures, set up display extraction sites, and distributed information about the various uses of peat among rural museums and schools. The launch of a regular journal *Vestnik Torfianogo Dela* in 1914 further indicated how much the administration wanted to turn peat extraction into a popular exercise. Apart from advocating peat as a fuel, these efforts also popularized other uses of peat, including its suitability as bedding material for livestock and a means to enrich the soil.[65]

Rural self-governing bodies (zemstvos) contributed to this campaign as well. Established in 1864 as part of Alexander II's Great Reforms to involve local actors in the governance of the countryside, zemstvos

63 For a survey on imperial agrarian reforms, see Carol S. Leonard, *Agrarian Reform in Russia: The Road from Serfdom* (Cambridge: Cambridge University Press, 2011), 25–61.

64 Yanni Kotsonis, *Making Peasants Backward: Agricultural Cooperatives and the Agrarian Question in Russia, 1861–1914* (New York: St. Martin's Press, 1999).

65 GUZiZ, *Obzor deiatel'nosti za 1913 god*, XVII, 160–161.

played a principal role in providing education, infrastructure, and health services to the rural population. In the decades before the First World War, agricultural experts working in rural self-government were crucially involved in promoting what they understood to be progressive modes of farming.[66] As part of this broader effort, zemstvos actively propagated peat extraction as a means to secure peasants' supplies of fuel and animal bedding. Their requests for land surveyors or financial assistance to drain wetlands and set up peat extraction sites reached the Turf Master's Division from all over European Russia.[67] In their appeals to the central authorities, zemstvo representatives often echoed the widespread sentiment that peasants needed assistance to gain the best utility from their land. That most peasants had no interest in extracting peat was a common complaint. "The peat is lying in vain (*vtune*). It is not brought anywhere and not used for anything by the local population," lamented a zemstvo delegate from the province of Vladimir in 1910. In light of the rapidly shrinking forest cover, he continued, instructing peasants to extract and use peat would serve the rural population and the region at large.[68]

While state officials and zemstvo employees saw peat extraction as part of a long-term strategy to develop the rural economy, peasants tended to extract it out of concrete necessity, most importantly, when they did not have enough firewood. In his 1896 survey on peat soils in Russia, botanist and soil scientist Gavriil A. Tanfilev mentioned that, in parts of the provinces of Kursk, Orlov, Tula, Tambov, and Ryazan, peasants burned peat in their homes.[69] Tatyana Tolstaya, Lev Tolstoy's oldest daughter, noted in her diary how she visited a village near her father's estate Yasnaya Polyana in 1891. The peasant huts she entered were so cold that Tolstaya could see her own breath. A local woman explained that they used peat to heat their homes. Firewood was hard to find, and only wealthier peasants could afford it. Others, the woman went on, burned

[66] Ilya V. Gerasimov, *Modernism and Public Reform in Late Imperial Russia: Rural Professionals and Self-Organization 1905–1930* (Houndmills: Palgrave Macmillan, 2009).

[67] E.g. Letter from the Zubstovskaia District *Zemstvo*, Perm' province, (1910), RGIA f. 425, op. 1, d. 35, l. 30.

[68] Letter from the Gorokhovetskaia District *Zemstvo*, Vladimir province, (1911), RGIA A f. 425, op. 1, d. 38, ll. 2–3.

[69] G. I. Tanfil'ev, "Torf," in *Proizdvoditel'nye sily Rossii: Kratkaia kharakteristika razlichnykh otraslei truda – sootvetstvenno klassifikatsii vystavki* (Saint Petersburg: Ministerstvo Finansov, 1896), 29.

potato leaves or dung. Tolstaya also learnt that peasants struggled to incorporate peat into their way of life. Peat fires were hard to light and, once lit, did not burn hot enough to bake bread.[70]

From the perspective of the modernizing elites, reluctance, disinterest in, or outright refusal to burn peat fuel only proved the rural population's ignorance. The same was true for what the elites considered poor and undeveloped extraction methods. Turning peat into fuel required a lot of experience. Industrial extraction sites therefore usually hired entire groups of seasonal workers (usually peasants) who traveled to the peat works every summer. Peasants, who turned to peat during an acute fuel shortage, were often less familiar with the best extraction techniques. State officials complained that they cut the peat "without any system" or "in very inadequate ways,"[71] allegations which sounded similar to imperial elites' disapproving comments about how peasants mined their land for minerals.[72] Peasant peat extraction, the experts stressed, lacked the necessary care. Digging with simple spades, the rural population often failed to create quarries with straight boundaries, as the manuals that began to be published at the time advised. Even worse, while some peat was left "wasted" in the ground, peasant practices created a landscape full of holes, which prevented productive agricultural after-use.[73]

The First World War further accentuated the difference between peasants' situational interest in peat and elites' ideas about what constituted a "correct" use of peat deposits. Facing severe fuel shortages, rural and urban administrations from all over European Russia approached the Ministry of Agriculture, requesting financial and practical aid to set up peat works. Such pleas arrived, for instance, from provincial or district zemstvos in Novgorod, Tver, Penza, Poltava, Pskov, or Simbirsk, as well as from the self-governing bodies of cities as large as Petrograd, Moscow, Ekaterinburg, or Kazan.[74] At times, peasants even approached the Ministry of Agriculture single-handedly, wishing to lease state-owned land for peat extraction or get advice on how they could exploit

[70] Tat'iana Sukhotina-Tolstaia, Diary, November 10, 1891, Prozhito, https://corpus.prozhito.org/note/32534. Accessed March 18, 2025.

[71] "Opisanie issledovannoi torfiannoi zalezhi Krivoe Boloto" (between 1900 and 1902), RGIA f. 425, op. 1, d. 21, ll. 89–90; "O rabotakh Torfmeisterskoi Chasti v 1914–1916 gg.," RGIA f. 425, op. 1, d. 43, l. 203.

[72] Pravilova, *A Public Empire*, 81.

[73] Vikhliaev, "Kratkii Ocherk," 268–270.

[74] List of rural and urban administrative bodies that contacted the Turf Master's Division between 1914 and 1916, RGIA f. 425, op. 1, d. 43, l. 139.

communal peatlands.[75] Ivan Vikhliaev noticed the rising interest in peat early on. Speaking to a conference on fuel matters in November 1915, he expressed concern that peasants had begun to "rip up" their wetlands. An outspoken advocate of popularizing peat extraction, Vikhliaev was sympathetic toward the situation of the rural population. Still, the uncoordinated exploitation of wetlands threatened to do more harm than good: "Such an irrational extraction of the peatlands will demand considerable expenses in the future to turn them into agricultural land. These expenses might be much bigger than the gain from the extracted peat."[76] Echoing the widely shared opinion that peasants were incapable of treating their environments responsibly, Vikhliaev's words embodied the tension between elitist and peasant perceptions of peat.[77] While extracting and burning peat represented a response to an emergency for much of the rural population, experts like Vikhliaev would have liked to see peat extraction for domestic needs embedded within a larger strategy of modern land management.

RESOURCES AND THE MAKING OF THE SOVIET STATE

A few years later, the administration Vikhliaev had served was no more. The 1917 February Revolution and the Bolshevik takeover in the same year toppled the social and political order that had formed over three centuries of Romanov rule. The Soviet state that emerged from the ruins of the Russian Empire married the revolutionary aspirations of Vladimir Lenin and his entourage with the violence of the Civil War period. At the same time, it relied considerably on experts and institutions from the Tsarist period that now served the Bolsheviks' efforts to build the Soviet state. These dialectics of persistence and transformation clearly showed in the early Soviet administration's approach to nature. The nationalization of land and resources immediately after the October Revolution rendered the state the most crucial authority over Russia's natural wealth. In light of a fuel crisis that grew more serious by the day, administrative attempts to map, quantify, and exploit peatlands continued and accelerated under

75 "O rabotakh Torfmeisterskoi chasti v 1914–1916 gg.," RGIA f. 425, op. 1, d. 43, ll. 14–15, l. 75.

76 *Trudy soveshchanii po podmoskovnomu ugliu i torfu, sozvannykh v Moskve na 20–22 noiabria 1915 goda Moskovskim upolnomochennym predsedatelia Osobogo soveshchaniia po toplivu* (Moscow: Biuro ob-edinennykh tekhn. organizatsii, 1916), 89.

77 On foresters' perception of peasant resource-use practices, see Brain, *Song of the Forest*, 23–28.

Soviet power. Renamed Peat Office (*Otdel torfa*), the Turf Master's Division became a permanent unit within the People's Commissariat of Agriculture (Narkomzem). Within months following the Bolshevik takeover, the Peat Office got tied to the new leaders' vision of transforming Russia's energy system. In the Spring of 1918, Ivan Vikhliaev advised the revolutionary government on peat extraction and the prospects of mobilizing peat to generate electric power. Peat fuel, which the imperial government had stimulated though without ever making it a priority, was now seen as essential to realize the promise of the revolution.[78]

From the point of view of the Soviet leadership, all peat extraction required administrative oversight. A testimony of the state's new position as the owner of all resources, this perspective also reflected widespread concerns about spontaneous resource extraction as it occurred during the Civil War, when massive tree-felling by peasants and competing state organizations caused a dramatic decline of forests in European Russia.[79] Against the background of the destruction, peat extraction once again appeared as a means to alleviate the pressure on wood resources. Narkomzem officials argued that "replacing firewood with peat in the everyday life of the rural population will serve as the most valid guarantee for the preservation of the rapidly shrinking forests."[80] At the same time, they contended that the broader adoption of peat fuel would necessitate close monitoring to prevent the uncontrolled exploitation of peatlands driven by acute need.[81] Soviet legislation reflected this desire to expand and simultaneously control local extraction. The 1922 Decree on Peatlands legally enshrined the state's authority over these environments, while a follow-up decree in 1925 distinguished between peatlands of "local" and "general state significance" (*mestnogo* vs *obshchegosudarstvennogo znacheniia*),[82] an approach

78 Vikhliaev's memoirs, RGAĖ f. 356, op. 1, d. 173, ll. 1–58; "Protokol zasedanii GOĖLRO" (1920), RGAĖ f. 5208, op. 1, d. 2, ll. 1–2, 9–10. On the role of peat in the early Soviet electrification campaign, see Chapter 2.

79 Brian Bonhomme, "A Revolution in the Forests? Forest Conservation in Soviet Russia, 1917–1925," *Environmental History* 7, no. 3 (2002): 411–434; Brain, *The Song of the Forest*, 63–67.

80 "Poiasnitel'naia zapiska k dekretu o torfianykh bolotakh" (1922), RGAĖ f. 478, op. 7, d. 1098, l. 4.

81 Ibid.

82 "O torfiankyh bolotakh" (May 17, 1922) and "O razdelenii torfianykh bolot na imeiushchie obshchegosudarstvennoe i mestnoe znachenie" (22 May 1925). Both available at https://docs.historyrussia.org. Accessed March 18, 2025.

which had also informed the 1923 Forest Code.[83] Under these new regulations, peatlands "of general state significance" were designated for exploitation by central organizations or enterprises. In contrast, those "of local significance" were meant to be used by the rural population for their own needs or by enterprises subject to local rather than central authorities. Combining imperial practice with the principles of the New Economic Policy, the decrees also stipulated that the state would impose rental charges for the use of peatlands.[84] As one official commented, the country's vast peat deposits could be turned into an "inexhaustible source of our state income."[85]

Peasant and state ideas about the value and uses of peatlands never fully converged. Even though more people in the countryside adopted peat fuel during the Great War and the extended revolutionary period, there was little enthusiasm for it. When supplied with sufficient wood to keep themselves warm and make food, peasants used peatlands primarily as pastures and hay meadows. As an internal Narkomzem document noted in 1922, "in many places, the interest in peat must still be brought to life."[86] Despite a campaign to instigate demand for peat fuel and peat-based agricultural inputs in the villages, which resembled similar initiatives from the late imperial period, peasant peat use fell short of expectations.[87] The Soviet administration abandoned rental charges for personal extraction in 1926, underscoring both the government's commitment to supporting rural households and the undeniable fact that peat remained uncommon as a domestic fuel. Even so, the state insisted on controlling how peatlands would be exploited. Official regulations prescribed that peasants had to cut the peat in orderly quarries instead of digging holes without draining the land beforehand. The administration also determined how much peat a rural household could claim and instructed that, to extract peat, peasants needed to set up peat collectives

83 Brain, *Song of the Forest*, 71.

84 "Dekret Soveta Narodnykh Komissarov o torfiankyh bolotakh" (1922), RGAĖ f. 478, op. 7, d. 1098, l. 62.

85 Note from the Head of the Division for Land Reclamation (1922), RGAĖ f. 478, op. 7, d. 1098, l. 1.

86 "Poiasnitel'naia zapiska k dekretu o torfianykh bolotakh" (1922), RGAĖ f. 478, op. 7, d. 1098, l. 6.

87 Report from a peat expert in the region of Riazan' (1924), RGAĖ, f. 478, op. 7, d. 2684, ll. 3–4. On the early Soviet peat campaign, see also Nikiforova, "'Useless Fossils'," 79–81.

(*tovarishchestva*).[88] In 1928, the Soviet peat expert P. V. Taneev noted with some delight that rural peat works expanded. In his view, however, still far too few peasants gained "utility" from their wetlands, using them mainly to make hay or to collect berries, brushwood, or moss to caulk their huts, "completely unaware that there is peat in the swamp from which one can make fuel or animal bedding."[89]

The tension between the state's developmentalist aspirations and local practices continued to haunt the authorities in the decades to come. The Soviet administration, although actively promoting peat as a fuel, fertilizer, and animal bedding, insisted on having complete oversight over how and by whom it was extracted. Returning to the established view that rural people understood little about peat, officials lamented its "predatory exploitation" through unsanctioned and careless "hole-digging," and demanded that local authorities regulated, recorded, and policed all extraction. Collective and state farms were to present a plan and officially request permission to dig up peat, even when they used their own land for this purpose. They were also prohibited from giving peat to consumers who were not formally entitled to it.[90] State organs had trouble enforcing these rules and often ignored their breach. Still, the mere fact that they existed meant that using peat in times of need, historically a common peasant practice during fuel shortages, presented a potential criminal offence.[91]

Meanwhile, state efforts to explore and map peatlands took an unprecedented scale. This was due, in particular, to the adoption of the 1921 GOĖLRO-plan, which firmly established peat as a material foundation of the nascent Soviet power industry and made knowledge about relevant peat deposits indispensable for the state.[92] Taking stock of peat available for extraction fell within the responsibilities of Narkomzem, which supervised the exploitation of what was now called the "state peat fund"

88 "Pravila razrabotki mestnym krest'ianskim naseleniem torfianykh zalezhei iz sostava gosudarstvennogo torfianogo fonda, dlia udovletvoreniia sobstvennykh nuzhd v toplive, podstilke i udobrenii" (1926), GARF, f. A-259, op. 10b, d. 512, ll. 10–12, 29–33.

89 "Sel'skie torforazrabotki," *Torfianoe delo* 3, no. 5 (1926), 114; Draft article "Kak krest'ianinu samomu uznat', est' li torf v bolote i na chto on prigoden" (1928), RGAĖ f. 9458, op. 1, d. 16, l. 1.

90 "Postanoveleniia i rasporiazheniia Pravitel'stva po torfu" (1943), GARF f. A-412, op. 1, d. 4, ll. 2, 31, 36, 42, 57; "Prikaz Glavtorfa 3 Avgusta 1942," RGAĖ f. 7969, op. 2, d. 5, ll. 162–163.

91 "Pravila i instruktsii po kontroliu za izspol'zovaniem torfianykh mestorozhdenii," GARF f. A-412, op. 1, d. 13, ll. 35–82.

92 Chapter 2 discusses the role of peat in Soviet electrification in greater detail.

(*gosudarstvennyi torfianoi fond*). In the early 1940s, after two decades of work, the ministry published comprehensive regional peatland inventories. The volume for the region of Moscow alone extended over almost 200 pages filled with tables that classified peatlands according to the volume and chemical properties of their peat.[93] The Soviet administration also introduced a "passport" (*passport*) for relevant peat deposits. Accounting for features such as the distance from transportation routes and vegetation, the document provided information on the depth, ash content, calorific value, and humidity of the peat – categories which clearly reflected an extractivist interest.[94] For decades an essential instrument of Soviet peatland inventory, the peatland passport signified the apex of a resource-making process that had begun some 200 years before. A few decades into Soviet power, the homogenizing gaze of the modern state had unmistakably reached the waterlogged environments of central Russia.[95]

* * *

In the twentieth century, Eurasia's peatlands were widely recognized as objects of scientific inquiry and resources for an industrializing economy. Imperial elites had employed social techniques of abstraction, simplification, and valuation since the Enlightenment period, constructing peatlands as promising environments whose value, at least in part, derived from the matter they stored beneath the surface. The Tsarist administration was profoundly involved in the mental and physical appropriation of Russia's peatlands, providing the knowledge, infrastructure, and financial means to map and explore them. The 1917 Revolution solidified the state's role in the exploration, governance, and exploitation of Eurasia's peatlands. Early Soviet experts and institutions built on and intensified existing approaches to making wetlands legible and continued the imperial tradition of treating them as potential sources of fiscal revenue. They also framed the rural population's uses of these landscapes as backward, unsystematic, and irrational. Still, the continued and increased state

93 *Torfianoi fond RSFSR: Torfianye bolota Moskovskoi oblasti* (Moscow: Narkomzem, 1940).

94 *Torfianoi fond RSFSR: Postanovleniia, polozheniia i itogi rabot po gosudarstvennomu torfianomu fondu* (Moscow: Narkomzem, 1945), 41–52; A. A. Sirin and T. Iu. Minaeva, eds., *Torfianye bolota Rossii: K analizu otraslevoi informatsii* (Moscow: GEOS, 2001), 45-46.

95 James C. Scott, *Seeing Like a State: How Certain Schemes to Improve the Human Condition Have Failed* (New Haven, CT: Yale University Press, 1998).

interest in peatlands under Soviet power brought some important changes. The access to and use of wetland environments now became matters of bureaucratic regulation and control to a much larger degree than ever before.

Alexander Solzhenitsyn's short story "Matryona's House" vividly depicts the practical challenges that the state's extended power over nature could create for people in the countryside. Set in the west of the Vladimir region shortly after Stalin died in 1953, when industrial peat extraction was already well established in the Soviet economy, the story revolves around Matryona, an older woman who takes a school-teacher and former prison convict as a lodger into her house. Matryona's life is difficult. Unable to keep up her work in the Kolkhoz, she gets caught up in an endless struggle with Soviet bureaucracy when applying for a state pension. Matryona grows her own food and forages for berries. Like other women from the area, she uses the summer months to secretly gather peat from the surrounding peatlands, which are officially controlled by a state-owned peat trust: "Just as people had once stolen wood from the landlord, now they scrounged peat from the Trust." Ending this practice was beyond the trusts' powers:

> What could the Peat Trust do? They weren't allowed enough staff to post watchmen all over the bogs. ... Now and again, at random intervals, they would send out a patrol and catch the women as they returned to the village. The women would drop their sacks and scatter. Sometimes too, when an informer gave them a tip-off, they would make a house-to-house search, compile lists of people caught hoarding illegal peat and threaten them with prosecution. The women would stop pilfering it for a while, but winter was coming on and drove them to it again – at night this time, and with sledges.[96]

When Solzhenitsyn wrote this story, peat theft had been a trope in Soviet discourse for quite some time. During the Second World War, a travelling unit of *Pravda* produced a satirical poster for the peat workers of Shatura, depicting men and women stealing peat from an open fuel train while the guard is asleep. Unlike Solzhenitsyn's story, which presented Matryona's act of pilfering with empathy, the poster was designed as a warning. Taking sides with the state, it called for the punishment of fuel thieves.[97]

[96] Alexander Solzhenitsyn, *Matryona's House and Other Stories*. Translated by Michael Glenny (Harmondsworth: Penguin Books, 1974), 18–19. On Solzhenitsyn's story and peat in early Soviet culture more generally, see Robert Bird, "The Poetics of Peat in Soviet Literary and Visual Culture, 1918–1959," *Slavic Review* 70, no. 3 (Fall 2011): 591–614.

[97] "K otvetu!," Pravda – Vyezdnaia redaktsiia, *Okna, plakaty, karikatury* (Shatura, 1944), no pagination.

Despite their different ways of looking at people and peat, both the image and the story hint at what the Soviet administration's extensive powers in the making and exploitation of resources meant for those living at the margins of Russia's rising fossil economy. Emanating from the state's role as the highest authority over the country's natural wealth were new boundaries between people and peat – and these prioritized industrial interests while marginalizing those who considered peat a survival resource.

2

Mobilization

In January 1920, engineer Gleb M. Krzhizhanovskii published a *Pravda* article stressing the importance of peat for the nascent Soviet state. Not only was Russia the country with the greatest peat reserves on earth, as Krzhizhanovskii pointed out, but these deposits were also very favorably located, particularly in the area of Petersburg (sic!) and in the Central Industrial Region around Moscow, where physical proximity to places of high industrial demand made peat an "ultra-local fuel" (*ul'tra-mestnoe toplivo*). Expanding peat fuel generation, particularly by mechanizing the extremely labor-intensive extraction process that required tens of thousands of seasonal workers every year, would considerably improve energy supplies in industrial and urban centers and reinvigorate Russia's war-ridden economy.[1]

Mobilizing the swamps promised much more than providing heat energy for companies and private users. It would help usher Russia into a new era: the age of electricity. In a follow-up article, Krzhizhanovskii emphasized the significance of what contemporaries called "local fuel" for a more fundamental shift in the country's energy regime. Regional power stations would exploit the flow of rivers and set the energy of brown coal basins and peatlands in motion to propel industrial development and make transport more efficient. Power utilities would also improve living conditions and increase food production in the countryside. For Krzhizhanovskii, electricity was not a value-neutral technology but a crucial building block of the new political order with its promises of

[1] G. M. Krzhizhanovskii, "Torf i krizis topliva," *Pravda* (January 10, 1920), 2–3.

social justice and political emancipation. In fact, electricity was no less than the energy of the revolution itself: "The age of steam was the age of the bourgeoisie. Steam is now being replaced by electricity, which is forging the rule of the proletariat."[2]

Krzhizhanovskii was not only a highly-regarded engineer, he was also an old friend of Lenin's. The two had first met in the 1890s when Lenin joined the revolutionary circles that Krzhizhanovskii had been involved in while a student in Saint Petersburg. Krzhizhanovskii's 1920 *Pravda* articles emerged from a series of meetings between himself and Lenin in December 1919, during which Krzhizhanovskii elaborated on the underappreciated energetic potential of locally available fuels, particularly peat. Not only did Lenin thereafter encourage his friend to write about the topic, he also actively intervened in the editorial process to ensure Krzhizhanovskii's ideas would be accessible to a broader public audience.[3]

Amidst the raging Civil War, Lenin had good reason to be excited about, of all things, peat. Fuel shortages had been a pressing issue in Russia for years, a problem that got much worse after the collapse of the Tsarist regime. Depictions of how desperately people in the cities searched for anything they could burn have featured prominently in contemporary sources and historical accounts of the revolutionary period. In the territories under Bolshevik control, getting hold of coal, oil, and even timber was almost impossible. Hospitals, museums, and other state institutions in the city of Moscow closed or remained unheated, as trees, furniture, and fences ended up in people's stoves.[4] Driving Lenin's involvement in the publication of Krzhizhanovskii's articles were his deep concern about the fuel crisis and his vision to place energy at the heart of socialist state-building. Power, for him, clearly had a material component. In February 1920, Lenin appointed Krzhizhanovskii to chair the State Commission for the Electrification of Russia (GOĖLRO). In the following year,

[2] G. M. Krzhizhanovskii, "Zadachi ėlektrifikatsii promyshlennosti," *Pravda* (January 30, 1920): 1.

[3] Alex Cummins, "The Road to NEP, the State Commission for the Electrification of Russia (GOELRO): A Study in Technology, Mobilisation, and Economic Planning," PhD Dissertation, University of Maryland (1988), 66–74; Jonathan Coopersmith, *The Electrification of Russia, 1880–1926* (Ithaca: Cornell University Press, 1992), 157.

[4] *Time of Troubles: The Diary of Iurii Vladimirovich Got'e. Moscow. July 8, 1917 to July 23, 1922.* Translated, edited and introduced by Terence Emmons (Princeton: Princeton University Press, 1988), 84, 89, 267, 278; S. A. Smith, *Russia in Revolution: An Empire in Crisis, 1890 to 1928* (Oxford: Oxford University Press, 2017), 218.

Krzhizhanovskii became the first head of Gosplan, the State Planning Committee, a clear indicator of the extent to which electrification and economic planning were entangled in the early Soviet state.[5]

The GOĖLRO forged a long-term relationship between power generation and resource extraction in central Russia, an aspect that has remained at the margins of historical scholarship. Works about Soviet electrification tend to focus on the interplay of technology, ideology, and expertise in the GOĖLRO. Others explore the cultural fascination with electricity during the early years of Soviet power when energetic metaphors permeated the political language and inspired artists and writers.[6] However, while literature and mass media imbued electricity with a mystic, immaterial aura, the planners in the GOĖLRO commission never considered electricity an immaterial force which would instantly bring comfort and progress. On the contrary, they had no illusion that generating electric power was impossible without primary energy carriers, be they water, coal, or peat. Most of their work involved gathering statistics documenting the role of various energy sources before the 1917 Revolution. The GOĖLRO's experts also created comprehensive resource inventories. Given that the planners faced an economy in disarray with a transport system damaged by years of war and revolution, the geographical distribution of resources was crucial for how they envisaged the spatial organization of the nascent Soviet economy and its power infrastructure. Thinking about the GOĖLRO, we must therefore center the "matters of power" and highlight their dual role as products of a longer resource-making process during which the country's elites appropriated Russia's natural riches for the benefit of an industrializing economy and as material structures shaping the geography and the workings of the power industry for decades to come.

Taking seriously the historical and geographical approach of the GOĖLRO commission, this chapter situates the mobilization of peat deposits for the power industry within a longer trajectory of industrial growth and resource use in central Russia. Industrial enterprises exploited

[5] The GOĖLRO is often considered the beginning of Soviet economic planning. See Heiko Haumann, *Beginn der Planwirtschaft: Elektrifizierung, Wirtschaftsplanung und gesellschaftliche Entwicklung Sowjetrusslands 1917–1921* (Düsseldorf: Bertelsmann, 1974).

[6] Coopersmith, *Electrification of Russia*; Anindita Banerjee, "Electricity: Science Fiction and Modernity in Early Twentieth-Century Russia," *Science Fiction Studies* 30, no. 1 (2003): 49–71; Richard Stites, *Revolutionary Dreams: Utopian Vision and Experimental Life in the Russian Revolution* (New York: Oxford University Press, 1989), 48–50.

the extensive peat deposits in the provinces of Moscow and Vladimir since the late nineteenth century. The fuel crisis of the First World War and the extended revolutionary period further enhanced the appreciation for "local" fuels. These developments sparked ambitious plans to convert peat into electric power that materialized with the construction of several peat-based power utilities in the first decade of Soviet power. From the point of view of central Russia's peatlands, the GOĖLRO intensified and accelerated already existing extractivist practices. At the same time, electrification made these landscapes part of a spatially integrated energy system which converted peat from a "local" fuel into a form of energy serving users whose interaction with peatlands was mediated by generators, transmission lines, and light switches. Notions of what was distant and what was close were constantly redefined as novel sources and forms of energy came to play an increasingly important role in Russia's rising fossil economy and the futures which contemporary actors imagined for it.

PEAT AND THE RISE OF RUSSIA'S INDUSTRIAL METABOLISM

To understand why Krzhizhanovskii was so optimistic about peat as an energy source, it is important to consider how industrial development had configured the exploitation and use of resources in Russia before the 1917 Revolution. Historians have foregrounded the role of energy as an essential driver of economic growth since the eighteenth century when the "subterranean forest" began to play an ever-increasing role in fueling industries, first in Great Britain and then in other parts of Europe and the United States. Industrialization is commonly regarded as a transition from a somatic (or organic) to a mineral (fossil fuel-based) energy regime.[7] However, the rise of energy-dense fossil fuels was neither inevitable nor uncontested. In Britain, where large industries relied on water wheels well into the nineteenth century, the nature of the property regime and market competition prevented a long-term reliance on water-based energy systems even when water was cheaper than coal.[8] Neither did the rise of industrial economies entail a straightforward substitution of mineral for organic energy sources. Instead, it is more accurate to conceptualize the energy

[7] Rolf Peter Sieferle, *The Subterranean Forest: Energy Systems and the Industrial Revolution*, translated by Michael P. Osman (Cambridge: The White Horse Press, 2001); Wrigley, *Energy and the English Industrial Revolution*.

[8] Malm, *Fossil Capital*, 114–120.

history of industrializing economies as a continuous process of intensification and additions affecting all available energy sources, not just fossil ones. Though their relative, and over time also their absolute, importance was shrinking, forests, animals, and human bodies retained their roles as suppliers of energy even where industries developed at a high pace.[9]

Industrial expansion brought far-reaching socio-metabolic changes in Russia as well. In the decades before the 1917 Revolution, industry, construction, transportation, and communication contributed a growing share to the national income.[10] Historians have examined the social, political, and cultural impact of Russia's industrial take-off. However, the material premises of growth, although recognized as significant factors enabling and driving industrial development in other parts of the world, have received only scarce attention. Considering that steam engines, which relied on energy-dense combustibles, were crucial in most sectors facilitating Eurasian growth in the late nineteenth century, the secondary role of energy and resources in the literature is a striking gap which obfuscates that economic growth, as much as it transformed society, also forged new entanglements between people and nature in the Russian Empire. Between 1887 and 1913, fuel use in industry and transportation grew by 292 percent as fossil energy sources slowly began to dominate overall fuel use in these sectors. The "mineralization" (*mineralizatsiia*) of Russia's fuel budget, as contemporary economist Lev B. Kafengauz called this process, was driven most notably by the skyrocketing increase in oil and coal use. At the same time, timber use declined in relative and, since 1908, absolute terms in both transportation and industry, though it remained the most important source of heat in private households, particularly in the countryside.[11]

These developments created new links between the various regions of the Russian Empire and changed the spatial organization of the late

9 Iñaki Iriarte-Goñi and María-Isabel Ayuda, "Not Only Subterranean Forests: Wood Consumption and Economic Development in Britain (1850–1938)," *Ecological Economics* 77 (2012): 176–184; Paul Warde, "Firewood Consumption and Energy Transition: A Survey of Sources, Methods and Explanations in Europe and North America," *Historia Agraria* no. 77 (2019): 7–32; On Barak, *Powering Empire: How Coal Made the Middle East and Sparked Global Carbonization* (Oakland: University of California Press, 2020), 8–11 and subsequent chapters on water, animals, and humans in environments shaped by the extraction, transport, and consumption of coal.

10 Paul R. Gregory, *Russian National Income, 1885–1913* (Cambridge: Cambridge University Press, 1982), 134.

11 Calculations based on Kafengauz, *Ėvoliutsiia promyshlennogo proizvodstva*, 30, 75, 130.

imperial economy. The Donets Basin coal mines and Baku oil fields emerged as the most important sites of fuel extraction, their role as suppliers of energy carriers becoming firmly anchored in the late imperial imagination. From the 1880s, trains and boats transported growing amounts of petroleum to the industrial hotspots of the Russian Empire, turning rivers and rails into axes of mobility not only for people and manufactured goods but also for the energy carriers that fueled industrial production.[12] The coal from the Donets Basin was consumed mainly by metallurgical companies in Ukraine, which ranked among the largest businesses in the Russian Empire. As the prime energy source for the expanding railway network, coal also underpinned the dynamic integration of the imperial space in the late nineteenth and early twentieth century.[13] This energy geography allowed for a massive expansion of industry and transport, but it also created new vulnerabilities. Industries and railways could operate only as long as fuel supplies were steady and reliable. The First World War would show that this was not self-evident.

But Russia's rising fossil economy also rested upon the mobilization of less iconic fuels. From the late nineteenth century, the peatlands of central Russia began to serve as fuel deposits for industrial enterprises. Scientists and noble landowners had experimented with peat as a source of heat since the eighteenth century. Yet, as long as the sunshine captured in the trunks of trees proved sufficient to fuel their commercial ambitions, peat extraction remained a relatively rare exercise. Lomonosov's vision to harness the energy of Russia's swamps turned into reality only in the second half of the nineteenth century, when the empire embarked on a trajectory of sustained economic growth. By the late 1850s, no more than 40 peatlands close to the city of Moscow had become extraction sites.[14]

[12] Even though the oil industry became oriented toward exports in the late imperial period, much of the oil from the South Caucasus was used in transport and industry within the Russian Empire. Sara C. Brinegar, *Power and the Politics of Oil in the Soviet South Caucasus: Periphery Unbound, 1920–1929* (London: Bloomsburg Academic, 2024), 12. On the cultural and economic importance of oil, see Jane Costlow, "Seeing Oil: Isaak Levitan and the Industrial Volga," in Catherine Evtuhov, Julia Lajus and David Moon, eds., *Thinking Russia's History Environmentally* (New York: Berghahn, 2023), 74–97; Nat Moser, *Oil and the Economy of Russia: From the Late-Tsarist to the Post-Soviet Period* (London: Routledge, 2018), 13–16; Rehschuh, *Aufstieg zur Energiemacht*, 32–40.

[13] Kafengauz, *Ėvoliutsiia promyshlennogo proizvodstva*, 24–31, 130; Volodymyr Kulikov and Martin Kragh, "Big Business in the Russian Empire: A European Perspective," *Business History* 61, no. 2 (2019): 299–321.

[14] Nikitinskii, *Torf i ego razrabotka*, 7–14.

In the following decades, industrial producers paid increasing attention to the bountiful peatlands near important manufacturing sites and along railway tracks to satiate their businesses' demand for fuel. Although the energy density of peat was low when compared to coal or oil, commercial elites incrementally began using it, particularly where peat deposits were located near manufacturing sites. The temporary character of peat works and the availability of seasonal labor in the countryside allowed factories to incorporate peat extraction into their running business. Between the late 1880s and the First World War, peat use in industry and transport increased by 188 percent.[15]

Most of the commercial demand for peat stemmed from the booming industrial districts of the Vladimir and Moscow provinces. As one of the most dynamic sectors in the late Russian Empire, the textile industry was crucial in the energetic mobilization of central Russia's wetlands. The All-Russia Art and Industry Exhibition in Moscow informed visitors in 1882 that the Morozovs' Nikolskaia Company in the very west of the Vladimir province, at that time the empire's largest textile mill, exploited around 1,300 desiatinas of state-owned peatlands which the company rented from the treasury. In the previous year, 5,227 men and women, which amounted to no less than a fifth of the corporation's total of 25,800 workers, had dug up peat to power the mill.[16] Over subsequent years, the role of peat in the company's operations would only increase. In 1895, peat accounted for more than 30 percent of the fuel used in the Nikolskaia factory.[17] The Morozov mills were no exception. By 1900, peat contributed around 20 percent of all fuel used in Russian textile enterprises and more than a third in the cotton mills of the Moscow and Vladimir provinces.[18] For most people at the time, however, peat remained invisible. While the fabrics these mills produced contained at least some tangible traces of their origin in the cultivated environments of distant cotton fields, peat's presence in creating these materials remained entirely hidden. Not surprisingly, the environments of central Russia's

[15] Calculations based on Kafengauz, *Ėvoliutsiia promyshlennogo proizvodstva*, 30, 75, 130.

[16] *Illiustrirovannoe opisanie Vserossiiskoi khudozhestvenno-promyshlennoi vystavki v Moskve 1882 g.* (Saint Petersburg: Tipografiia Ėduarda Goppe, 1882), 43, 46, 158.

[17] I. V. Potkina, *Na olimpe delovogo uspekha: Nikol'skaia manufaktura Morozovykh* (Moscow: Izdatel'stvo Glavarkhiva Moskvy, 2004), 81–82.

[18] Kafengauz, *Ėvoliutsiia promyshlennogo proizvodstva*, 30–31; N. P. Langovoi, "Ocherk manufakturnoi promyshlennosti," in *Fabrichno-zavodskaia promyshlennost' i torgovlia Rossii* (Saint Petersburg: Ministerstvo Finansov, 1896), 16.

wetlands rarely featured in cultural and historical representations of the late imperial textile industry.

This evidence is important for how we think about the nature of economic growth in the late Russian Empire. As in other parts of the world, industrial expansion rested upon a significant increase in energy consumption. This involved the exploitation of coal and oil deposits in the empire's southern and western parts as well as the mobilization of the rich carbon stocks of central Russia's temperate peatlands. Focusing on peat fuel reveals the diversity of the late imperial economy and its fuel sources. Scholars increasingly recognize the relevance of regional perspectives in attempting to grasp the immense range of lived experiences and identities within the empire's boundaries.[19] The changing patterns of energy use in the post-emancipation period only underscore how valuable this "regional paradigm" is for scholars of economic history.[20] Peat was marginal in the aggregated energy budget, but its consumption could reach substantial levels in certain parts of the empire. The massive peatlands of the Meshchera Lowlands stretching to the east of Moscow offered the thriving industrial enterprises in the area a fuel they could access cheaply, promptly, and without being exposed to the erratic fluctuation of fuel prices. Introducing an environmental perspective to the history of Eurasian growth requires that we take the importance of the region as a unit of analysis seriously while recognizing the fluid nature of regional identities and boundaries over time.

But the significance of peat extraction also transcended the regional level. Environmental historian John McNeill has argued that the Industrial Revolution created "ecological teleconnections" that bound distant corners of the world together in a global process of economic expansion and intensifying resource use.[21] Textile mills in central Russia,

[19] Evtuhov, *Portrait of a Russian Province*; David Saunders, "Regional Diversity in the Later Russian Empire," *Transactions of the Royal Historical Society* 10 (2000): 143–163; Susan Smith-Peter, "How to Write a Region: Local and Regional Historiography," *Kritika: Explorations in Russian and Eurasian History* 5, no. 3 (2004): 527–542; Katherine Pickering Antonova, "Discovering Russian Regions: Fruits of the Archival Turn in Imperial Russian History," *Kritika: Explorations in Russian and Eurasian History* 21, no. 4 (2020): 877–892.

[20] Katherine Pickering Antonova, "Regional'naia paradigma ėkonomicheskogo razvitiia v Rossii XIX veka," in Ekaterina Boltunova and Willard Sunderland, eds., *Regiony Rossiiskoi imperii: Identichnost', reprezentatsiia, (na)znachenie.* (Moscow: NLO, 2021), 52–74.

[21] John R. McNeill, "Cheap Energy and Ecological Teleconnections of the Industrial Revolution, 1780–1920," *Environmental History* 24, no. 3 (2019): 492–503.

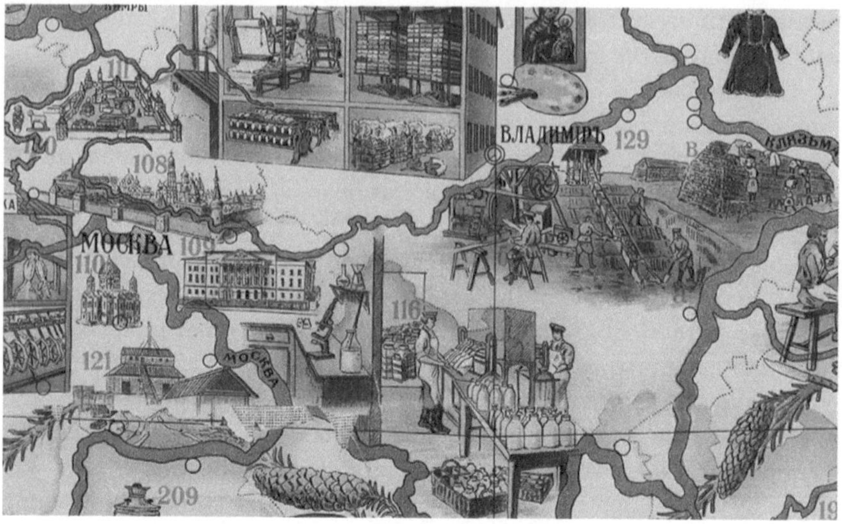

FIGURE 2.1 Map extract showing peat extraction as an important economic activity in central Russia. *Source:* Illiustrirovannaia karta Evropeiskoi Rossii. Sostavlena M. A. Pavlovnoi pod redaktsiei A. S. Barkova i S. G. Grigor'eva (Moscow: Izdanie Tv-a I. N. Kushnerev & Co, 1896).

which contributed to the rise of capitalist labor relations already before the end of serfdom, relied on such ecological teleconnections as well. Their extensive demand for fibers transformed distant social and physical geographies, particularly in the cotton-growing regions of Central Asia. But these factories also imprinted on environments and workers' lived experiences in places not directly linked to cotton farming or the weaving and coloring of fabric.[22] In the late nineteenth century, much of the heat required to convert fibers into clothes originated from peatlands in central Russia, making them one of the many factors that rendered Central Asia's cotton fields profitable. Focusing on seemingly marginal places such as the Meshchera wetlands in the borderlands of the Vladimir and Moscow provinces showcases the myriad ways in which industrial development integrated places, people, and things which have, thus far, rarely been regarded as interconnected elements of a shared political ecology (Figure 2.1). It was his awareness of the role that peat had played in

[22] Alison K. Smith, "A Microhistory of the Global Empire of Cotton: Ivanovo, the 'Russian Manchester'," *Past and Present* 244, no. 1 (2019): 163–193; Maya K. Peterson, *Pipe Dreams: Water and Empire in Central Asia's Aral Sea Basin* (Cambridge: Cambridge University Press, 2019), 134–142.

Russia's expanding industrial metabolism that would excite Krzhizhanovskii when he began to project the future of the economy as the head of the GOĖLRO.

PEAT FUEL DURING RUSSIA'S CONTINUUM OF CRISIS

The First World War quickly undermined the spatial organization of the late imperial fuel economy. With the collapse of coal imports, the German occupation of the Dombrowa coal basin, increased energy demand by the defense industry, and an overloaded railway system, fuel shortages became a constant problem in the empire's industrial hotspots.[23] Manufacturers whose business operations had previously relied primarily on coal and oil increasingly looked at fuels from local sources. In October 1914, just a few months after Germany declared war on the Russian Empire, the directorate of a metal rolling plant approached the Ministry of Trade and Industry, asking for permission to buy 300 desiatinas of a peatland south of Petrograd. In light of dramatically rising timber prices and increasing difficulties in obtaining coal and oil, the company planned to switch their steam engines to peat. The war exacerbated a problem that the mill had experienced for quite some time. According to the company's account, the price of wood fuel had more than doubled in previous years as the forests in the area had come under pressure from extensive logging.[24] Other industries became interested in peat as an emergency fuel as well. In January 1916, for instance, the 1886 Company for Electric Lightning, Russia's leading supplier of electric energy, sought to rent a peatland close to Petrograd because fuel shortages threatened one of their power plants and, as a result, the electricity supply of the capital's munition factories.[25] Responding to requests like these, various parts of the imperial administration became involved in making peatlands available for extraction. The First World War made the mobilization of these environments appear an urgent necessity. Locally extracted peat now promised to sustain production, profit, and the empire's defensive forces.

While absolute peat extraction for commercial purposes declined during the war because most companies lacked workers and machinery,

[23] Peter Gatrell, *Russia's First World War: A Social and Economic History* (Harlow: Pearson Education Limited, 2005), 108–119.

[24] Letters from the Ministry of Commerce and Industry and from the Siverskie Metal Works Company (October 1914), RGIA f. 425, op. 1, d. 48, ll. 208–209.

[25] Letter from the Company to the Chief Administration for Resettlement and Agriculture (January 1916), RGIA f. 425, op. 1, d. 43, l. 172.

the conflict initiated a shift toward the consolidation of peat extraction as an economic sector of its own. Peatlands had been an object of state governance and industrial interest since the second half of the nineteenth century, when the Tsarist administration explored profitable ways to govern state-owned lands and central Russian businesses began exploiting peat deposits close to their factories. Yet, unlike coal or oil extraction, which attracted substantial amounts of capital from domestic and foreign investors, peat extraction remained mostly attached to individual enterprises that used peat but did not trade it, while attempts to turn peat extraction into a business in its own right rarely succeeded. During the fuel emergency of the war, peat commissions and committees mushroomed within state institutions and public associations. The Central War-Industry Committee, a lobby organization of industrialists, and the Special Council for Fuel, which the Tsar instituted in response to the economic challenges inflicted by the war, both had special units to support and expand peat extraction. The Russian Association for Trade and Industry similarly created a section to deal with peat. In 1915, the Ministry of Agriculture initiated a Peat Committee, which involved representatives of the Ministries of the Interior, Transport, Finance, Trade and Industry, as well as delegates from various agricultural associations and regional branches of the Central War Industry Committee. Emblematic of the mobilization of society under the auspices of the Tsarist state, these new platforms became crucially involved in the collective effort to exploit Russia's peat deposits during the war.[26]

In practice, these attempts boiled down to a myriad of painstaking negotiations and efforts to address the practical challenges of extracting and transporting peat. Could relevant parts of the central administration be convinced that peat held crucial importance for the empire's war efforts and that the interests of peat extraction ranked similarly to those of coal mining? Would the Ministry of Transport agree to introduce special rail tariffs for peat shipments? How could enterprises get hold of relevant machinery after imports had become nearly impossible? Closely involved in the Peat Committee, Ivan Vikhliaev, the head of the Turf Master's Division in the Ministry of Agriculture, spent lengthy meetings examining applications from private landowners, industrial companies,

[26] "1-oe Zasedanie Torfianogo Komiteta," *Vestnik Torfianogo Dela* no. 3–4 (1915), 382–388. On Russia's parastatal complex during the First World War, see Peter Holquist, *Making War, Forging Revolution: Russia's Continuum of Crisis 1914–1921* (Cambridge, MA: Harvard University Press, 2002), 20–21.

and local administrations seeking financial support to set up peat extraction works or briquette factories. As the war wore on, the members of the Peat Committee grew frustrated at their limited ability to regulate and mobilize Russia's peat resources effectively. In 1916, they agreed to lobby for legislation reflecting the rising demand for peat. Some even proposed that such legislation include the state's prerogative to expropriate privately owned land if a drainage canal had to be built across private property.[27] The Tsarist government never passed such a law and failed to develop a coherent strategy to govern the empire's peat deposits. Though partly a reflection of the secondary role of peat fuel, this had to do primarily with the imperial state's limited success in mobilizing nature for the war more broadly. Attempts to make coal supplies more reliable through state intervention also showed meagre results, even though coal ranked highly among the government's priorities.[28] Still, even though it did not lead anywhere, the Peat Committee's decision to press for a law on peat highlights a rising consensus according to which the state would coordinate the governance of Russia's environments for the sake of the common good, however vaguely defined. This idea would surface in a new guise after the October Revolution when private ownership of resources was abandoned and peatlands became part of a resource pool managed and exploited by government agencies.

The fuel crisis of the war opened up new imaginaries of what the future of energy might be. In the turbulent summer of 1917, Ivan Vikhliaev and his staff in the Turf Master's Division, which had overseen the mapping and exploitation of peatlands since the 1880s, drafted a budget for the coming 1918 year. It was an extraordinary period for such an ordinary task. In the cities, various political factions debated Russia's political future, as peasant unrest spread through rural regions, and deserting soldiers undermined the army's ability to continue the war. The archival record of the Turf Master's Division bears little trace of these dramatic developments. Keeping up with their routine, Vikhliaev and his colleagues estimated the treasury's expected income from leasing state-owned wetlands and projected how many and which peatlands they planned to assess. Yet, unlike in previous years, their budget also entailed broader reflections on the role of peat fuel in the time ahead. The experience of the war clearly dictated their thoughts. Whenever peace would come, they

[27] "Iz deiatel'nosti Torfianogo Komiteta," *Vestnik Torfianogo Dela* no. 3–4 (1915): 383–388; no. 1 (1916): 85–110; no. 2 (1916): 248–302; no. 3–4 (1916): 395–445.

[28] Pravilova, *A Public Empire*, 273–274; Gatrell, *Russia's First World War*, 96–97.

wrote, there would be no return to the old days, at least regarding fuel. Forests would hardly be able to satisfy the fuel demands of a recovering economy, while prices for coal and oil were unlikely to fall. Due to their convenient location, peatlands in central and north-western Russia would become critical reservoirs of industrial fuel. As the officials made clear, the advantages of peat were not only tied to the price of other fuels. At least when compared to wood, whose share in overall energy use had increased again during the war, appropriating peat deposits allowed for the extraction of more energy from a given surface area, because the energy density of peat was higher than that of wood.[29]

Presumably, 1918 and subsequent years looked far different than Vikhliaev and his staff had imagined when drafting the budget. The Bolshevik takeover in October 1917 ended the fragile division of power between the Provisional Government and the Soviets that had emerged after the abdication of Tsar Nicholas II. During the violent battle over the territories and peoples of the dissolving Russian Empire, access to fuel became a matter not just of economic recovery but, indeed, of political survival for the Bolshevik government. While the urban population struggled to keep their houses warm, factories whose production had been severely hit already during the world war faced permanent fuel shortages. Following the Treaty of Brest-Litovsk, which separated Ukraine, the Baltic provinces, Finland, and most of Belorussia from Russia, and during the fights over the oil fields of Baku, there was little chance that sufficient amounts of coal or oil would reach the industrial centers held by the revolutionary forces. Struggles over railway lines and the disruption of railway traffic due to fuel shortages further undermined the circulation of resources during the Civil War.[30] Facing the collapse of the late imperial energy geography, the revolutionary government had no choice but to make the mobilization of local fuels a political priority.

Peat now received attention at the highest levels of the emerging Soviet state. The nationalization of land immediately after the Bolshevik takeover relegated control over natural resources to the central administration. The Decree on Peat Fuel from April 21, 1918 declared the expansion of peat extraction a matter of utmost urgency and subjugated all related

[29] "Smeta raskhodov po torfodobyvaniiu na 1918 god," RGIA f. 425, op. 1, d. 43, ll. 339–345.

[30] On the fuel crisis during the Civil War, see Coopersmith, *Electrification of Russia*, 130–137. On the course and geographies of the Civil War, see Smith, *Russia in Revolution*, 161–183.

activities to the Supreme Council of the National Economy (VSNKh), the highest economic authority in the new administration. VSNKh's regional organs received the right to claim and distribute all available peat fuel. Continuing the imperial government's attempts to organize the war economy, the Soviet leadership also established a Chief Peat Committee to oversee the mobilization of peat resources in regions under Bolshevik control.[31] Forest historians have stressed that the aspiration of state-led resource management was unattainable as long as Russia's political future remained contested.[32] This was true as well for peatlands. Difficulties in recruiting workers and supplying them with food impeded any significant expansion of extraction, while peasants' spontaneous turn to peat fuel left many peatlands degraded.[33] Still, the first months of Soviet power proved significant as they cemented the state's role as the principal authority over Russia's natural resources. At the same time, they marked the gradual consolidation of peat extraction as an economic sector on its own. Unlike in previous decades, when the state's dealings with peatlands had revolved primarily around managing the treasury's lands, peat was now widely regarded as a resource of strategic importance. This new recognition of the value of peat would ultimately pave the way for its long-term role within the Soviet energy system.

Tracing how industrial and administrative elites in Russia reconsidered the value of peat and, in turn, attempted to expand its extraction during the First World War and the Civil War underscores the benefits of applying an environmental lens to the scholarship surrounding this period.[34] The Great War left a deep mark on ecosystems around the world. While trenches and military waste desolated landscapes in the immediate combat zones, regions far from the theaters of war turned into extractive frontiers. The war also increased state control over resources

31 "Dekret Soveta Narodnykh Komissarov 'O Glavnom Torfianom Komitete'" and "Dekret Soveta Narodnykh Komissarov 'O razrabotke torfianogo topliva'" (April 1918). Available at https://docs.historyrussia.org. Accessed March 19, 2025.

32 Bonhomme, "Revolution in the Forests?," 420–423; Brain, *Song of the Forest*, 63–67.

33 *Otchet Glavnogo Torfianogo Komiteta: Dobycha torfa v 1919 g. i snabzhenie prodovol'stviem torfianikov za tot zhe period* (Moscow: VSNKh, 1920), 3–4. See Chapter 1, on peasants' use of peatlands during the revolutionary period.

34 Andy Bruno, "What Does It Mean to Liberate a Land? Toward an Environmental History of the Russian Revolution," in Christopher Read, Peter Waldron, and Adele Lindenmeyr, eds., *Russia's Home Front in War and Revolution, 1914–22*, Book 3: National Disintegration (Bloomington: Slavica Publishers, 2018), 157–177; Jennifer Keating, *On Arid Ground: Political Ecologies of Empire in Tsarist Central Asia* (Oxford: Oxford University Press, 2022), 181–209.

and expanded the reach of exploitative practices that deeply interfered with the reproductive processes of nature.[35] Like the expansion of tin mining in Southeast Asia or the mobilization of nitrate reserves in Latin America, the recognition of – and attempt to exploit – peatlands as nationally relevant resources reflected a global trend toward the intensification of resource use that had been triggered by industrialization, gained in pace during the war, and undergirded extractive practices during the twentieth century. Although most efforts to mobilize Russia's peatlands during the war failed to meet contemporaries' aspirations, they still laid the foundation for the massive expansion of peat extraction in the decades to come. Historian Peter Holquist argued that the years from 1914 to 1921 were fundamental in shaping the future relationship between state and society, as the institutions and practices of mobilization emerging during the war became institutionalized within the early Soviet state.[36] But the legacies of this period also extended to the nonhuman world. The experience of fuel shortages made industrial and technical elites appreciate peat as a resource that would play a prominent role in the post-war economy. Affecting the ideas and practices of resource use for decades to come, Russia's Continuum of Crisis had lasting political as well as ecological implications.

INTO THE ELECTRIC AGE

It was its recognition as a resource for the electric age which eventually sparked the extraction of peat on a massive scale. Celebrated as a new social and economic force around the world, electricity had gained a foothold in Russia during the 1880s. Its use spread only slowly though. While a small but vibrant community of engineers engaged with technological advances abroad and developed electrotechnical equipment, the Tsarist government remained largely unmoved by the radiant visions attached to electricity elsewhere. Unlike the railways, which administrators regarded as a crucial tool to modernize and govern the empire, electricity received little state support. As a result, electric facilities developed primarily due to the initiative of factory owners who sought to power their production. In addition, a few commercial suppliers, such

[35] Tait Keller, "The Ecological Edges of Belligerency: Toward a Global Environmental History of the First World War," *Annales. Histoire, Sciences Sociales (English Edition)* 71, no. 1 (2016): 61–78.

[36] Holquist, *Making War, Forging Revolution*, 3.

as the 1886 Company, a stockholding company financed largely with Siemens-Halske capital, provided public lighting and power for industrial facilities in several larger cities such as Saint Petersburg or Moscow. Until the late imperial period, most utilities relied on foreign technology and capital, remained small in size, and served only a few customers. In 1910, 95 percent of the empire's cities still had no electricity.[37]

During the First World War, plans to expand electricity generation coalesced with calls for mobilizing locally available energy sources. The war proved crucial for the generation of electricity in many countries. In Germany and the United States, for instance, governments funded power infrastructure to accommodate the increased demand for electric energy. The war years also brought about a shift toward the construction of larger and increasingly interconnected utilities, a legacy that would underpin power network design for decades to come.[38] Although there was no notable expansion of power utilities in the Russian Empire, technical elites increasingly gravitated toward the idea that electricity constituted a fundamental technology. Facing mounting fuel shortages, engineers and members of government departments began to embrace regional power utilities as tools to foster industrial development.[39] Fuel shortages also brought home the fact that the spatial order of the empire's economy, where urban and industrial centers depended on resource deposits located hundreds – if not thousands – of kilometers away, made many industrial businesses extremely vulnerable to any disruption in the transport system. Converting fuels with a lower energy density, such as brown coal or peat, into electricity promised a solution to the intertwined transportation and fuel crises.

Krzhizhanovskii was among the first to articulate a clear vision for a regionalized power industry that relied on what he and others called "local" fuels. In November 1915, the Special Council of Fuel Supply summoned a conference to discuss ways to use the largely untapped coal and peat reserves in the province of Moscow. One of the speakers at the event, Krzhizhanovskii emphasized peat's significance as an energy source in the Central Industrial Region. For him, peat was not just a means to bolster Russia's industrial metabolism and its demand for combustibles

37 Coopersmith, *Electrification of Russia*, 47; Haumann, *Beginn der Planwirtschaft*, 13–20; S. N. Simonov, *Razvitie ėlektroėnergetiki Rossiiskoi imperii: Predystoriia GOĖLRO* (Moscow: Universitet Dmitriia Pozharskogo, 2018).

38 Thomas P. Hughes, *Networks of Power: Electrification in Western Society, 1880–1930* (Baltimore: The Johns Hopkins University Press, 1983), 285–292.

39 Coopersmith, *Electrification of Russia*, 114–120.

during war times but a vehicle for a more fundamental shift in the wider economy: its transition into the electric age. The engineer presented a list of more than 30 already existing factory-based electric facilities in central Russia and underscored the technical feasibility of turning peat into power. Krzhizhanovskii also stressed that much could be gained from developing not just individual plants serving a single customer but a network of power providers. It was at this conference that the future architect of the GOĖLRO plan first projected a system of regional power facilities which would make central Russian businesses and cities independent from long-distance fuel shipments. In this imaginary, wetlands were nothing less than the natural foundation of the future economy: "In the peatlands of the Central Industrial Region, we have colossal stocks of local fuel for many dozens of regional stations, which for a century may distribute life-giving light and kinetic energy through their power lines. Reality imperiously pushes us in this direction, and we will have to pay dearly if we don't listen to this call."[40]

The engineer's optimism was grounded in experience. As the commercial director of Ėlektroperedacha, Russia's earliest peat-based regional power plant, Krzhizhanovskii had first-hand knowledge about using peat to generate electric energy. Operating since shortly before the beginning of the war, the station was located close to the town of Bogorodsk (known since 1929 as Noginsk) in the Meshchera Lowland, where textile and other industrial mills had been using peat fuel since the late nineteenth century. Ėlektroperedacha demonstrated that an integrated power infrastructure held clear advantages over individual industrial facilities. In late 1914, the station provided electricity to illuminate the streets of Bogorodsk, as more than 200 consumers, among them leading textile mills, were connected to the station's transmission lines. From 1915 onward, the city of Moscow received power from Ėlektroperedacha as well. During the war, the station provided as much as 20 percent of Moscow's electrical energy.[41] In the summer of 1917, when Vikhliaev reflected on the post-war future of the economy, he referred to Ėlektroperedacha as an example that proved the tremendous potential of peat.[42]

[40] G. M. Krzhizhanovskii, "Oblastnye ėlektricheskie stantsii na torfe i ikh znachenie dlia tsentral'nogo promyshlennogo raiona," in *Trudy Soveshchanii po Podmoskovnomu ugliu i torfu*, 126.

[41] Coopersmith, *Electrification of Russia*, 81–82; Simonov, *Razvitie ėlektroėnergetiki*, 168–172.

[42] "Smeta raskhodov po torfodobyvaniiu na 1918 god," RGIA f. 425, op. 1, d. 43, ll. 39ob–34o.

Those advocating for expanding the empire's power infrastructure tended to entertain technocratic visions of an interventionist state, which gained much traction during the First World War as governments in belligerent countries became more active in allocating resources and defining production targets in strategic sectors. In the interwar period, the European Left and Right alike sympathized with the idea of economic regulation at the national level.[43] Technocratic state intervention gained traction in the Russian Empire as well, as the Tsarist government tried – mostly unsuccessfully – to address war-related bottlenecks in strategic sectors such as food and fuel.[44] Despite their different political loyalties, members of the officialdom and the critically minded educated class agreed that the state and technical elites should play a central role in organizing Russia's economy and society during wartimes and beyond. Projections about the future of electricity reflected these ideas. Engineers saw economic regulation and electrification as two closely intertwined components of the national economy. Ėrnest Bukhgeim, for whom electric energy was an absolute economic necessity, called for a shift from private, and particularly foreign, capital toward increased state involvement in the construction and operation of power utilities. Thermal engineer Vasilii Grinevetskii, whose works on post-war recovery anticipated early Soviet approaches to planning, equally highlighted the fundamental importance of electricity for the wider economy. Writing against the background of acute fuel shortages, both authors also highlighted the potential lying in low-calorific "local" fuels. Electric power, economic planning, and rational resource use now seemed intrinsically linked.[45]

The foundation of the GOĖLRO commission made state doctrine of the triad of electrification, planning, and resourcefulness, which engineers had formulated in the final years of the Tsarist regime. Involving more than 200 specialists, the commission embodied the technocratic spirit of

[43] Dirk van Laak, "Planung: Geschichte und Gegenwart des Vorgriffs auf die Zukunft," *Geschichte und Gesellschaft* 34, no. 3 (2008): 305–326; C. S. Maier, "Between Taylorism and Technocracy: European Ideologies and the Vision of Industrial Productivity in the 1920s," *Journal of Contemporary History* 5, no. 2 (1970): 27–61.

[44] Holquist, *Making War, Forging Revolution*, chapter 1. On the First World War as an important moment in the prehistory of Soviet economic planning see most recently Friedrich Asschenfeldt and Max Trecker, "From Ludendorff to Lenin? World War I and the Origins of Soviet Economic Planning," *Europe-Asia Studies* 76, no. 1 (2024): 10–28.

[45] Ė. O. Bukhgeim, *K ėkonomicheskomu osvobozhdeniiu Rossii putem ėlektrifikatsii ee territorii* (Moscow: S. P. Iakovlev, 1915); Coopersmith, *Electrification of Russia*, 116, 140–142.

the early Soviet government. Most members had no links with the Bolshevik party, but they shared the revolutionaries' enthusiasm for technological progress and endorsed the opportunity to get involved as system-builders.[46] Under Krzhizanovskii's direction, they gathered extensive statistical records about the historical evolution and current state of economic affairs. They also undertook detailed surveys of relevant natural resources. Based on this information, the commission planned a network of interconnected power utilities and defined priorities for the future development of the economy at large. In December 1920, Lenin and Krzhizhanovskii presented the work of the GOĖLRO, a document several hundred pages in length, to the Eighth Congress of the Soviets. A carefully choreographed display of the plan, during which numerous lightbulbs illuminated a large map of Eurasia, underscored what both men expressed in their speeches: Electrification represented not just a means of recovery but a vehicle for transformation. As Lenin declared famously, electrification was now the "second party program."[47]

Geography lay at the heart of the GOĖLRO. Brought together to reimagine the economy, the planners made environmental determinants a central focus of their work. In contrast to the immaterial aura surrounding electricity in the contemporary cultural imagination, much of the commission's work revolved around matters of power in a literal sense. The massive fuel shortages of the preceding years featured prominently in their considerations. Without solving the fuel crisis, the introduction to the GOĖLRO report stated, one could not even think about a "rebirth" (*vozrozhdenie*) of the country.[48] Against this background, preventing shortages appeared as a matter, first and foremost, of using energy resources close to where they occurred instead of relying on the railways to carry them over long distances. Advocating a spatial distribution of industry that reflected the geography of natural resources, the plan divided Russia into economic macro-regions, for which it projected a

[46] Cummins, "Road to NEP," 119–132. On the role of pre-revolutionary engineers in the early Soviet state, see Susanne Schattenberg, *Stalins Ingenieure: Lebenswelten zwischen Technik und Terror in den 1930er Jahren* (Munich: Oldenbourg, 2002), 70–73.

[47] Coopersmith, *Electrification of Russia*, 174–178; Haumann, *Beginn der Planwirtschaft*, 167–177.

[48] *Plan ėlektrifikatsii RSFSR: Vvedenie k dokladu 8-mu S-ezdu Sovetov Gosudarstvennoi Komissii po Ėlektrifikatsii Rossii* (Moskva: Gosudarstvennoe tekhnicheskoe izdatel'stvo 1920), 25. One quantitative study states that "the Great War and Civil War induced the deepest crisis that Russia experienced in more than a century of modern economic growth." See Andrei Markevich and Mark Harrison, "Great War, Civil War, and Recovery: Russia's National Income, 1913 to 1928," *The Journal of Economic History* 71, no. 3 (2011): 672–703, 694.

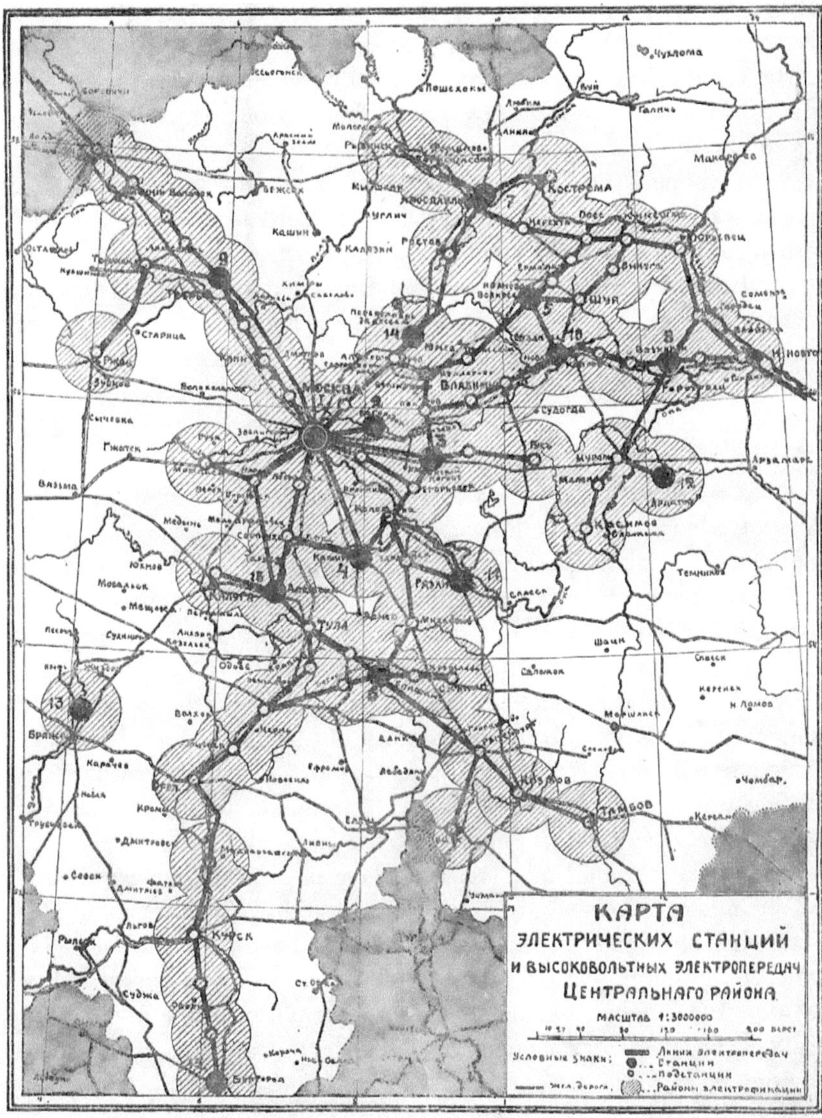

FIGURE 2.2 Map of electric power plants and transmission lines as envisioned by the GOĖLRO commission. *Source:* Nauchno-tekhnicheskii Otdel VSNKh, *Ėlektrifikatsiia Tsentral'nogo Raiona* (Moscow: Gosudarstvennoe tekhnicheskoe izdatel'stvo, 1920).

network of interconnected regional electrical stations fueled primarily by nearby energy sources (Figure 2.2). For the so-called Northern Region, which included the regions of Petrograd, Vitebsk, and Novgorod stretching to Murmansk, the members of the GOĖLRO commission assumed that hydropower would play a central role. Several of Petrograd's power plants were projected to run on peat and timber waste from sawmills. In the Central Industrial Region, most power plants would be based on low-calorific coal or peat to overcome the previous dependence on coal and oil brought from afar.[49]

Such emphasis on the spatialities and materialities of electricity networks was not, in itself, peculiarly Soviet. Plans to transport electricity instead of fuel also proliferated in other countries, as did suggestions to strengthen the state's role in promoting integrated power networks. In Mexico, for instance, the state became a major player in regulating and providing power during the mid-1920s.[50] In Germany, fuel shortages during the First World War and the ensuing Allied occupation of the Rhineland led to the construction of power utilities close to basins of low-calorific brown coal. Public interest in hydroelectric power increased as well. In the United States, social reformer and conservationist Gifford Pinchot advocated Giant Power, an ambitious scheme to improve the spatial and social reach of electricity in Pennsylvania through a network of interconnected power lines and large mine-mouth utilities.[51] Unlike Giant Power, which failed due to fierce resistance against, most notably, the degree of state regulation envisaged by its designers, the GOĖLRO left a visible trace on the Soviet power landscape. Not only did it establish the state as the principal electricity provider but, as the historian Felix Frey demonstrated, the GOĖLRO also enshrined "regional autarky" as a leading principle of the Soviet energy system for the decades to come. During the years of the First Five-Year Plan, the Stalinist leadership reiterated that mobilizing "local" fuels was central to expanding the power industry. It was only in the decades following the Second World War when the Soviet government emphasized the construction of large hydroelectric dams and later the exploitation of newly discovered oil and gas deposits in Siberia, that the energy regionalism of the revolutionary years gave way to efforts aimed at integrating the Soviet energy system on a national scale.[52]

Beyond spatial considerations, Krzhizhanovskii and the members of the GOĖLRO had another reason to favor using local fuels, particularly those of lower quality, in the power industry. According to the historian

49 *Plan ėlektrifikatsii RSFSR*, 206–214. 50 Vergara, *Fueling Mexico*, 169.

51 Hughes, *Networks of Power*, 297–313.

52 On "regional autarky" as an analytical concept and its relevance for Soviet energy history, see Frey, *Arktischer Heizraum*, 17–23.

Oscar Sanchez-Sibony, "interconnections with and reproductions of essential capitalist practices and institutions were always at the core of the Bolshevik project and must be incorporated in any explanation of its trajectory – even during its first decade."[53] The GOĖLRO is a clear case in point. Despite the Bolshevik promise of revolutionary economic change, the commission worked on the assumption that the Soviet state would remain closely entangled with international markets. Extrapolating from the central role of foreign capital and technology in the power industry during previous decades, the planners imagined that Russia's transition into the electric age was hinged upon imports of technology and know-how.[54] Germany and the United States would become the main players in the "conquest" (*zakhvat*) of the Russian market. The planners, eager to address the ongoing deficit in turbo generators, lamps, and other electrotechnical equipment through relevant imports, found this a promising prospect. Capitalist competition would benefit the socialist economy by allowing the Soviet state to negotiate advantageous trade agreements.[55]

Foreign currency was essential to make such a scenario work, but it was extremely scarce since the First World War had disrupted most international trade. Primary exports thus became an important building block in the architecture of the GOĖLRO, while the domestic use of lower-quality fuels was seen as a strategy to free up resources that were in demand internationally. As in previous decades, when imperial elites had advocated peat fuel for the sake of forest conservation, the value of peat was once again defined in terms of how it could replace goods that were in short supply or could be traded abroad. In his report to the GOĖLRO commission, Ivan Vikhliaev, previously the head of the Turf Master's Division within the imperial Ministry of Agriculture, stressed the benefits of using peat instead of timber. Apart from being essential for the reconstruction of the war-ridden country, timber should be saved for export "as the most real commodity (*naibolee real'nyi tovar*) in the near future," an argument also featuring in several later Soviet publications.[56]

The desire to attract foreign currency also informed the GOĖLRO's take on petroleum. A key motivation in the Bolshevik's struggle for Baku,

[53] Oscar Sanchez-Sibony, "Global Money and Bolshevik Authority: The NEP as the First Socialist Project," *Slavic Review* 78, no. 3 (2019): 694–716, 699.

[54] Coopersmith, *Electrification of Russia*, 185–186.

[55] *Plan ėlektrifikatsii RSFSR*, 20.

[56] "Materialy k voprosu ob ispol'zovanii torfianykh bolot v sviazi s ėlektrifikatsiei Rossii," RGAĖ f. 5208, op. 1, d. 59, l. 310b.; *Torfianoe delo v RSFSR i deiatel'nost' Torfianoi sektsii za period 1923–1924 gg.* (Moscow: Narkomzem, 1925), 5–6.

oil was critical to consolidating Soviet power in the South Caucasus.[57] Still, the authors of the GOĖLRO plan did not view oil as a vital component of the energy system they intended to build. Although they acknowledged that oil would retain some importance while the Soviet economy transitioned into the electric age, they saw the "centrifugal" and "decentralizing" nature of oil-powered motors as conflicting with their vision of a spatially integrated energy system. Instead, GOĖLRO planners, like the political actors in the South Caucasus, valued oil for its potential to generate foreign currency. By limiting domestic combustion and selling oil on international markets, they aimed to acquire urgently needed production equipment from abroad.[58] In light of these considerations, "regional autarky," with its emphasis on low-calorific energy carriers, appears as the flip side of the larger premise that the Soviet economy would, for the foreseeable future, develop interdependently with rather than in isolation from international markets.

NATURE'S PROMISE

Nourishing GOĖLRO's optimism about peat as an energy source that could fuel the power industry for decades, if not centuries, ahead were Cornucopian visions of abundance that had inspired the appropriation of peatlands since the eighteenth century. The planners were aware that peatlands existed all over the temperate parts of Eurasia; the question was, how could these environments be brought into productive use? To their amazement, the abundance of peatlands was not static but expanding over time, an idea grounded in the very nature of these landscapes. Peatlands are ecosystems in which living and dead organic materials, water, and inorganic substances constantly interact. Peat accumulates as waterlogged conditions prevent the decay of moss, grass, trees, or other vegetation. Contemporaries in nineteenth- and twentieth-century Russia were familiar with the fact that, in active peatlands, the build-up of matter is an ongoing process that, though hardly visible, distinguishes peat from older energy carriers, such as coal or oil. A Tsarist official commented as early as 1841 that there was "such a significant amount of peat in the province of Moscow that there is no need to fear its complete extermination." In line with the imperial government's conservationist concerns at the time, he portrayed peat as a renewable resource. Unlike

57 Brinegar, *Power and the Politics of Oil*, 21–46.

58 *Plan ėlektrifikatsii RSFSR*, 50–57; Rehschuh, *Aufstieg zur Energiemacht*, 41–53.

forests, which in some places were already at the brink of exhaustion, peat could recover from human use: "Due to its annual growth, the peat will continuously renew itself."[59]

No less excited about peatlands as places of constantly emerging matter, early Soviet planners liked to dwell on the idea that peat existed in such volumes that it would be foolish not to exploit it. Although they did not place peat into the same category of renewability as wind or water, they still emphasized that it was a peculiar resource because its continuous accumulation worked against – or delayed – depletion. In his 1920 *Pravda* article "Peat and the Fuel Crisis," Krzhizhanovskii celebrated the wetlands of central Russia as a resource that effectively took care of itself:

> A peatland is a living accumulator of heat energy (*zhizvoi akkumuliator teplovoi ėnergii*). The average annual growth of the peat layer is about 1,5 cm. If we extract billions of *pud* of peat, we will still not keep up with the natural growth of our peatlands. Our peat riches are truly non-depletable (*po-istine neizcherpaemy*).[60]

Given Krzhizhanovskii's leading role in the GOĖLRO commission, it is not surprising that the notion of peatlands as self-replenishing resources found its way into the commission's final report as well: "With peat, we could cover our entire fuel demand without disrupting its natural renewal cycle (*prirodnyi vozobnovitel'nyi tsikl*)."[61] Leonid K. Ramzin, thermal engineer and another high-ranking GOĖLRO member, gestured in the same direction at the 1924 World Power Conference in London, where engineers from across the globe discussed the possibilities of transnational economic integration through interconnected energy systems. In his comprehensive report on the power resources of Russia, Ramzin stressed that "peat bogs are not dead deposits of fuel" and that they were capable of replenishing themselves when used for energetic purposes. In the European part of Russia, where peat extraction was most intense, "no exhaustion of peat resources takes place at present, but on the contrary their continuous growth and accumulation."[62] In peatlands, it seemed, nature was working for the human good.

59 "Ob otvrashchenii istrebleniia lesov upotrebleniem torfa vmesto drov," *Zhurnal Ministerstva Gosudarstvennykh Imushchestv* no. 1 (1841), 123.

60 G. M. Krzhizhanovskii, "Torf i krizis topliva," *Pravda* (January 10, 1920): 2.

61 *Plan ėlektrifikatsii RSFSR*, 40.

62 Leonid K. Ramzin, "The Power Resources of Russia," *The Transactions of the First World Power Conference 1924. Vol. 1 Power Resources of the World Available and Utilised* (London: Percy Lund Humphries & Co, 1925), 1223. On the conference, see

This image of dynamic abundance was more than a rhetorical device to justify the mobilization of Russia's peat stocks. At a time when preparing peat fuel entailed a long sequence of mostly manual tasks, it was hard to imagine that large-scale mechanized extraction would eventually leave many peatlands degraded and almost devoid of peat. Peatlands' continuous accumulation of matter made the depletion of peat appear even less likely. Historians have described the early Soviet electrification campaign as an example of Soviet Prometheanism, the quest to conquer and transform nature for the sake of progress.[63] There is no doubt that the architects of the GOĖLRO aimed to expand the amount of available electric energy through far-reaching interventions in the physical landscape. River damming, resource extraction, and the construction of power plants and transmission lines featured prominently in the plan. Posters and pamphlets translated these ambitions into memorable slogans and images. Yet the thinking of Krzhizhanovskii and other early Soviet planners went beyond a simplistic notion of nature as an object of human desire. Instead, they recognized that humans were always entangled with the material world rather than simply acting upon it. Whatever the economic goal, it would have to account for the workings of nonhuman nature. Even though the planners miscalculated the possible impact of their vision on wetland ecosystems, their attentiveness to natural cycles of renewal and decay is remarkable.

The defining feature of this "Marxist productivist ecology," as Daniela Russ succinctly labeled it, was the fusion of industrialism with a metabolic understanding of the economy. This approach emphasized the flows of matter and energy as fundamental components of all productive, consumptive, or reproductive activity. Within this framework, labor was described and assessed in energetic terms, that were applied to humans as well as plants, animals, rocks, and water.[64] The language of the GOĖLRO was imbued with productivist imagery of human and nonhuman labor. The final document spoke of the economy as a "working organism" (*rabotaiushchii organizm*) whose different parts were

Daniela Russ, "Speaking for the 'World Power Economy': Electricity, Energo-Materialist Economics, and the World Energy Council (1924–78)," *Journal of Global History* 15, no. 2 (2020): 311–329.

63 Gestwa, *Die Stalinschen Großbauten*, 56–58; Paul Josephson, "'Projects of the Century' in Soviet History: Large-Scale Technologies from Lenin to Gorbachev," *Technology and Culture* 36, no. 3 (1995): 519–559.

64 Daniela Russ, "Energetika: Gleb Krzhizhanovskii's Conception of the Nature–Society Metabolism," *Historical Materialism* 29, no. 2 (2021): 188–218.

"organically" (*organicheski*) bound to each other. It also referred to the working population as "living power" (*zhivaia sila*), suggesting that humans were part of a larger energetic whole which extended to the nonhuman world and needed to be governed as efficiently as possible.[65] Such ideas also permeated discussions of particular aspects of economic life. Suggesting that the "labor of the sun rays" (*rabota solnechnykh luchei*) encapsulated in the trunks of trees could be utilized much more rationally if electrical saws and sawmills substituted human labor in the woodworking industry, the GOĖLRO-plan placed humans within, and not above, a shared energy system which had its origins in the light and the heat of the sun.[66] In passages like these, the commission's materials are reminiscent of the holistic categories of Vladimir Vernadsky, whose works were among the first systematic attempts to place life on Earth, including humans, within a planetary framework.[67]

Bolshevik cosmology opened up a space for materialist ecology, but it did not invent it. The GOĖLRO's focus on energy and matter stood in a tradition of materialist thought, dating back to the middle of the nineteenth century when scientists and intellectuals from the Russian Empire began to engage with Justus Liebig's work. The German chemist highlighted the importance of nutrient exchanges between the soil, living organisms, and the atmosphere for the growth of plants, suggesting that chemical fertilizers could improve agricultural yields. Widely read among agriculturalists across Europe, Liebig's works also inspired Karl Marx's concept of the nature–society metabolism and his ecological criticism of modern farming, for long an omitted aspect of his intellectual legacy. Marx feared that a fissure between the food-producing countryside and food-consuming cities would disrupt nutrient flows and undermine the natural conditions of sustained plant growth. In late imperial Russia, revolutionary writers frequently framed their social and political criticism using soil metaphors.[68] Considerations of material flows resonated in the works of the GOĖLRO and beyond. In a report to the commission, Ivan

[65] *Plan ėlektrifikatsii RSFSR*, 13 [66] Ibid., 50.

[67] Giulia Rispoli and Jacques Grinevald, "Vladimir Vernadsky and the Co-evolution of the Biosphere, the Noosphere, and the Technosphere," *Technosphere Magazine*, June 20, 2018, https://technosphere-magazine.hkw.de/p/Vladimir-Vernadsky-and-the-Co-evolution-of-the-Biosphere-the-Noosphere-and-the-Technosphere-nuJGbW9KPxrREPxXxz95hr. Accessed March 18, 2025.

[68] Mieka Erley, *On Russian Soil: Myth and Materiality* (Ithaca: Northern Illinois University Press, 2021), 38–47. On Marx's concept of the metabolic rift, see Foster, "Marx's Theory of Metabolic Rift"; Saito, *Marx in the Anthropocene*, 23–28.

Vikhliaev, for instance, suggested using nitrogen, a by-product of peat-burning gas generators, as a fertilizer. This idea, which was reminiscent of materialist approaches in the soil sciences and reflective of contemporary anxieties about soil exhaustion, also spoke to a broader concern about waste reduction that featured prominently in early Soviet plans to integrate the power industry with other economic sectors.[69] Later in the twentieth century, such concerns would surface in the context of what the historian Elena Kochetkova called "Soviet industrially embedded ecology," the active search for ways to reconcile economic growth with a careful approach to nature.[70]

Even though the GOĖLRO plan promoted an agenda hinged upon expansive resource use, it carried important traits of progressive ecological thought. The plan combined an awareness that human agency could potentially cause a "metabolic rift" with an optimistic belief in science and technology as tools for organizing the nature–society metabolism in the most efficient ways.[71] Understanding economic growth in material and energetic rather than monetary terms, its authors foregrounded geography as a parameter which, at least in part, defined the avenues of Russia's economic future. They also stressed that industrial expansion was about mobilizing resources without undermining nature's productivity. For them, the shift from steam to electric power would mark a crucial step toward this goal. It promised to relieve pressure from the transportation system, one of the significant energy consumers, and to expand the energetic base of the economy without leaving a trace of environmental devastation. The planners' focus on hydropower, or "white coal" (*belyi ugol'*) as they called it, drew from such notions of efficiency as did their enthusiasm for peat as a "renewable" resource. Unlike in the pre-revolutionary period, when profit-driven enterprises had rapaciously (*khishchnicheski*) exploited Russia's natural riches, the relationship between the social and the natural worlds was set to improve. The country would depart from capitalism's exploitative approach to nature and move toward higher stages of development without losing sight of the material premises of production and reproduction. The idea of peat as a renewable energy source is clearly at odds with present-day

69 "Materialy k voprosu ob ispol'zovanii torfianykh bolot v sviazi s ėlektrifikatsiei Rossii," RGAĖ f. 5208, op. 1, d. 59, l. 33ob; Nikiforova, "'Useless Fossils'," 77–78.

70 For a broader discussion on waste and attempts to avoid it in the late Soviet period, see Elena Kochetkova, *The Green Power of Socialism: Wood, Forest, and the Making of Soviet Industrially Embedded Ecology* (Cambridge, MA: The MIT Press, 2024).

71 Russ, "Energetika."

understandings. Still, the GOĖLRO's conceptualizations of resources and their uses matter in the context of current efforts to think about the histories and futures of renewable energy. Conceptions of renewability have not been stable over time and assumed an important place in early Soviet energy imaginaries.[72]

EMBEDDING PEAT IN THE SOVIET ENERGY SYSTEM

The Shatura Regional Power Plant, located about 130 kilometers east of Moscow, exemplifies how fuel shortages, enthusiasm for electric energy, and visions of natural plenty and renewability coalesced in the early Soviet period, inspiring the creation of an energy landscape whose structural features continue to shape the region to this day. Built amidst the extensive peatlands of the Meshchera Lowland and surrounded by three lakes ideally suited to provide the station with cooling water, Shatura was one of the signature projects of the GOĖLRO. Its origins predated the revolution, though, as Robert Ė. Klasson, a patron of many Bolshevik engineers in the late imperial period and the initiator of Ėlektroperedacha, just 70 kilometers away, had first conceived a peat-fueled power station in the area in 1914. Construction eventually began after the Bolshevik takeover. A small power plant which mainly served as an experimental site started to operate in Shatura as early as 1920.[73] When the GOĖLRO commission began its work, the planners envisaged Shatura as a regional power hub that would supply the city of Moscow and the east of the Moscow province with electric energy. Only the Kashira Station, fueled by coal from the Moscow basin, was expected to surpass the capacity of 100,000 kW, which, according to the plan, Shatura would achieve in 1930. The cities of Ryazan, Ivanovo-Voznesensk, Yaroslavl, Nizhny Novgorod, and Vladimir were all to receive peat-based power plants of a similar size.[74]

These goals were ambitious. Dense forests and marshy ground abounded in the location that Klasson had identified as the future site of

[72] On renewability as an unstable concept, see Ute Hasenöhrl and Patrick Kupper, "Historicizing Renewables: Issues and Challenges," *History and Technology* 37, no. 4 (2021): 397–410. On the Soviet case, Russ, "'Socialism Is Not Just Built for a Hundred Years'."

[73] Simonov, *Razvitie ėlektroėnergetiki*, 270–271.

[74] *Ėlektrifikatsiia Tsentral'nogo Raiona. Sostavleno Gosudarstvennoi Kommissiei po Ėlektrifikatsii Rossii* (Moscow: Gosudarstvennoe Tekhnicheskoe Izdatel'stvo, 1920), 66–74.

the Shatura power station. From early 1918, workers arrived on foot or horseback to cut down trees and put in place some provisional infrastructure necessary to facilitate the upcoming building works. Most of them slept in dug-outs, tents, or quickly erected barracks. The remainder lodged with peasant families in surrounding villages.[75] The actual construction of the Shatura Power Station followed a familiar pattern in the history of Russian electrification – its dependence on expertise and technology from abroad. Foreign engineers and mechanics formed an essential part of the workforce. Shatura's boilers and generators were imported. Still, Shatura left an important mark on the later history of the Soviet power industry. Ivan I. Radchenko and Vladimir D. Kirpichnikov invented innovative ways to extract and burn peat on a large scale, which would subsequently be adopted in other regions. Alexander V. Vinter, Shatura's lead engineer, later headed the construction of the Dnepr Hydroelectric Station, another beacon of Soviet energy infrastructure.[76]

Emotions ran high when Shatura was ready to send electric power to consumers more than 100 kilometers away. In December 1925, leading government representatives, engineers, journalists, and foreign diplomats boarded a train in Moscow to attend the opening ceremony in Shatura. Leon Trotsky compared the new power station to a leap in time that liberated Russia from its deeply rooted historical backwardness: "Shatura, this is the twentieth century, and, maybe, even the twenty-first that has poured into the seventeenth century." The station was surrounded by swamps, but the masts carrying the wires to Moscow were "the same as in America," Trotsky stressed.[77] For him, this was a clear indicator that the revolution had set Russia on the right path to join the ranks of the most advanced nations at the time. Commenting on the opening of the Shatura Power Plant, journalist and writer Mikhail Kol'stov was enchanted by the alchemy of this "palace in the swamps," which transformed the "lowest-grade fuel," peat, into "the highest-grade energy," serving millions of people.[78]

Still, there was more to this story than Soviet myth-making. One of the first infrastructural projects accomplished under the new regime, Shatura

[75] "Vospominaniia o Shature," GARF, f. 7952, op. 3, d. 459, l. 6.

[76] Coopersmith, *Electrification of Russia*, 215, 246; A. V. Vinter, "Ot Shatury do DneproGĖSa," in *Sdelaem Rossiiu ėlektricheskoi: Sbornik vospominanii uchastnikov Komissii GOĖLRO i stroitelei pervykh ėlektrostantsii* (Moscow: Gosduarstvennoe ėnergeticheskoe izdatel'stvo, 1961), 32–38.

[77] "Torzhestvennoe otkrytie Shaturskoi ėlektrostantsii," *Pravda* (December 8, 1925): 2.

[78] Ibid.

soon played a critical role as a provider of electric power in one of the most industrialized regions of the early Soviet state. As the first advocates of peat-fired power stations had promised, electricity considerably extended the spatial reach of peat fuel. A few years after its opening, Shatura became part of Moscow's unified network of power stations (MOGĖS), a decision reflective of the centripetal tendencies embedded in the early Soviet power system.[79] In the second half of the 1920s, power demand in the wider Moscow region surpassed the capacity of the few existing power stations, forcing planners to enhance power generation capacities further. Shatura again turned into a construction site, as the station was extended to a size well beyond what the GOĖLRO-plan had initially projected. From the early 1930s, Shatura operated at a capacity of 130,000 kW, a significant increase in its ability to produce power and in the material demands placed on the surrounding wetlands from where the station drew its fuel.[80]

Most importantly, Shatura was not a singular case. Electrification made peat a staple resource of Russia's expanding fossil economy, turning wetlands near several cities into extraction sites. As the designers of the GOĖLRO had intended, peat-based power plants became a vital element in central Russia's power infrastructure. In 1928, such utilities accounted for over 40 percent of overall Soviet electricity generation. In 1940, they still contributed around 20 percent of the electric energy. Most of it went to industrial users.[81] Peat's significance in the early Soviet energy system was particularly evident in the regions of Moscow and Gorky (present-day Nizhny Novgorod), which became the most electrified part of the country. Rich peatlands supplied power stations of different sizes, the most important of which was the regional power station in Balakhna, near the city of Gorky. First commissioned in the 1920s, Balakhna's capacity reached a capacity of 204,000 kW in 1933, surpassing even Shatura. The early Soviet power infrastructure was similarly tied to peat fuel in the regions of Ivanovo, Yaroslavl, and Tver. In the Russian

79 "Spravka ob organizatsii mezhduvedemostvennoi komissii po vyrabotke uslovii peredachi MOGĖSu Shaturskoi i Kashirskoi ėlektrostantsii i ėlektrotoku ėlektrostantsii 'Krasnyi Oktiabr'" (1927), GARF, A-259, op. 11b, d. 1617, ll. 2, 5–7.

80 "Piatiletnii perspektivnyi plan ėlektrifikatsii Moskovskogo raiona (1926/27)," RGAĖ, f. 4372, op. 4, d. 145, ll. 1–21; *Shaturskaia GRĖS* (Moscow: Informėnergo, 1970), without page numbers.

81 Bodo Lochmann, "Vom GOELRO-Plan zum Energieprogramm: Zur Entwicklung der Elektroenergiewirtschaft der UdSSR seit den zwanziger Jahren," *Zeitschrift für Geschichtswissenschaft* 35, no. 10 (1987): 872–883, 881.

northwest, the Dubrovka station, which began operating in 1933, sent electricity from the peatlands stretching along the banks of the Neva River to the city of Leningrad.[82]

Peat increasingly played a role at the municipal level as well. When urban administrations from the regions of Ivanovo, Yaroslavl, Kostroma, and Vladimir complained in the early 1930s about forest depletion and their difficulties in acquiring fuel, the People's Commissariat of Communal Economy advised them to provide for themselves by extracting peat (*samozagotovka*). Narkomzem routinely rededicated land for this purpose. Mirroring the crucial role of cheap labor in preparing peat fuel, prison inmates were mobilized in at least one instance in the Leningrad region, where peat extraction for communal needs was the highest.[83] The extraction of peat for the municipal economy fit well with the larger energy strategy of the Stalinist state with its commitment to "regional autarky."[84] Although Krzhizhanovskii, the lead author of the GOĖLRO-plan, was sidelined after Stalin came to power, the Soviet regime remained committed to the plan's core principles.[85] The XVIII Party Congress in 1939 reiterated that the ambitious growth target, which the third Five-Year Plan set for the Soviet economy, required a considerable expansion of fuel production. Between 1937 and 1942, peat extraction was projected to grow by 206 percent, an increase which exceeded the intended growth rates for coal (181 percent) and oil (177 percent). Like in the days of the GOĖLRO, the rationale for expanding peat extraction was to limit the use of "distant" energy carriers for local needs.[86] In line with this ambition, an extensive campaign was afoot from the late 1930s to substitute peat and other local fuels for oil in municipal heat and power plants.[87] Taken individually, these efforts look like hectic

[82] Eric Thiel, "The Power Industry in the Soviet Union," *Economic Geography* 27, no. 2 (1951): 107–122.

[83] "Razvitie samozagotovok torfa" (1934), GARF f. A-314, op. 1, d. 8901, ll. 4–5; "Spravka o khode zagotovki torfa i zagotovki topliva dlia kommunal'nykh predpriatii v gorodakh Lenoblasti," GARF f. A-314, op. 1, d. 8901, l. 15. Chapter 3 discusses the role of cheap, including forced, labor at greater detail.

[84] Frey, *Arktischer Heizraum*, 100–106.

[85] Kendall E. Bailes, *Technology and Society under Lenin and Stalin. Origins of the Soviet Technical Intelligentsia, 1917–1941* (Princeton: Princeton University Press, 1978), 185.

[86] *XVIII S-ezd Vsesoiuznoi Kommunisticheskoi Partii (b) 10–21 marta 1939 g. Stenograficheskii otchet* (Moscow: OGIZ, 1939), 294, 652, 654.

[87] "Postanovlenie Soveta Narodnykh Komissarov RSFSR 'O khode vypolneniia plana dobychi mestnykh vidov topliva i obespecheniia na zimu toplivom mestnykh promyshlennykh i kommunal'nykh predpriiatii, shkol, bol'nits i naseleniia" (1940), GARF f. A-314, op. 1, d. 9113, ll. 30–33; "Po voprosu toplivosnabzheniia kommunal'nykh ėlektrostantsii

and uncoordinated attempts to respond to temporary fuel shortages. However, seen together, they reflect how normal it had become for administrators to think about peatlands as fuel deposits in the first instance. Half a century after imperial industries had begun to rely on the carbon stored in central Russia's wetlands, peat now featured prominently both in the economic imagination of Soviet bureaucrats as well as in the actual workings of the country's energy system.

The Second World War proved that the GOĖLRO's emphasis on local energy sources made a lot of sense. German occupation of the Donets coal basin and the Soviet dismantling and partial evacuation of the Caucasian petroleum industry severely disrupted fuel supplies. Once again, peat acquired significance as a remedy for war-related fuel shortages. Due to its role as a crucial electricity provider for the city of Moscow, the Shatura Power Plant attracted the attention of the German Luftwaffe, which produced a series of detailed aerial photographs of the station and adjacent extraction sites (Figure 2.3). Shatura endured the war undamaged and continued delivering electric power to the Soviet capital and its factories. And Shatura was no exception. Some 40 kilometers north of the city Gorky, the large Balakhna station provided power to the arms industry during the war, despite heavy bombing of the city. The situation looked different in regions closer to the frontlines. Recognizing the vital role of peat as a regional fuel, the German armed forces destroyed peat extraction sites in the Belarussian SSR and the Dubrovka power station just outside the besieged city of Leningrad. Due mainly to the efforts of civilians conscripted to extract peat from the nearby wetlands that had not fallen under German control, Leningrad's Red October power plant kept generating electricity during the war.[88] Deeply entrenched in the flows of matter and energy that enabled Soviet citizens to produce things and keep warm, peatlands experienced the war together with the people who had become dependent on them.

* * *

From the last decades of the nineteenth century, extensive parts of Eurasia's nature became entangled in the rhythms and logic of industrial

Narkomkhoza RSFSR v zimu 1940 – 1941gg.," GARF f. A-314, op. 1, d. 9113, 61–66; "O khode dobychi torfa po sisteme kommunal'nogo khoziaistva i oblastnykh, gorodskikh toplivnykh trestov" (1939), GARF f. A-314, op. 1, d. 396, ll. 16–19.

[88] Kopenkina, *Istoriia torfianogo dela*, 174–181; V. I. Markov, "Periody razvitiia torfianoi promyshlennosti Rossii," *Trudy Instorfa* 6, no. 59 (2012): 10–21.

FIGURE 2.3 Shatura Power Plant and surrounding peat extraction and peat storage sites. Aerial photograph taken by the German Luftwaffe in 1942. *Source:* Courtesy US National Archives (DT/TM-5/Russia, Schatura).

production. This trend reverberated strongly in central Russia's peatlands. The energy demands, first of industrial enterprises and later of power utilities, increasingly complemented plants, animals, the soil, and the elements as factors defining life in these environments. Focusing on peat reveals three important moments that should inform our understanding of the interplay between industrial expansion and energy since the late imperial period. First, low-calorific "local" fuels played a more significant role in Russia's rising fossil economy than previously assumed. Industrially driven fuel extraction, with all its social and ecological ramifications, was a reality in central Russia as much as in the better-known coal mining regions of the Donets basin or the oil fields of the South Caucasus. Second, the early Soviet electrification campaign reflected historical geographies of resource use and industrial development dating back to the second half of the nineteenth century. That the planners in the GOĖLRO commission believed local energy carriers would become a pillar of the Soviet energy system had much to do with the successful appropriation of such fuels in the late imperial period and the unsettling experience of the fuel crisis during the years of war and revolution. The GOĖLRO was undoubtedly tied to the Bolshevik experiment, but it also needs to be placed within a more extended temporal framework that accounts for both the economic trajectories it incorporated and its material legacies that shape central Russia's landscapes until the present day. Third, the fact that the GOĖLRO plan combined productivist and metabolic notions of economic development highlights the importance of moving beyond simplistic Prometheanist narratives about the relationship between nature and Soviet power.[89] Economic planning in early Soviet Russia drew on earlier traditions of materialism and geographical determinism in the Russian Left, fusing these with the imperatives of economic recovery and transformation that gained traction after the 1917 Revolution.[90]

But the forgotten history of peat in central Russia has broader lessons to teach. Because of its low energy intensity, peat is positioned somewhat awkwardly within the master narrative of the transition from wood to coal to oil and natural gas. Barely mentioned in most surveys of energy history, peat has been classified as a "traditional" source of heat. Its use in industrial societies is usually treated as a secondary phenomenon. The case of Russia challenges these views. Indeed, it was the rise of the fossil economy that prompted the appropriation and physical mobilization of peat for energetic purposes. Russia's resource geography and the elites' understanding of it provided a low-calorific energy carrier with a solid place in the industrializing economy. Although peat's share in overall energy use would shrink over time, its embeddedness in central Russia's energy system profoundly transformed wetland environments, as these became subject to massive industrial exploitation. Centering historical narratives on peat may help us appreciate nature as a factor that has shaped the social and spatial order of Russia's fossil economy. Peatlands played a crucial role in the country's move into the industrial era.

[89] On the "dualistic conception of nature," see Bruno, *Nature of Soviet Power*, 11–15.

[90] Mark Bassin, "Geographical Determinism in Fin-de-siècle Marxism: Georgii Plekhanov and the Environmental Basis of Russian History," *Annals of the Association of American Geographers* 82, no. 1 (1992): 3–22.

However, their contribution has been rendered invisible, first by technologies which turned peat into an industrial throughput and a power infrastructure which increased the distance between fuel extraction and energy consumption, and second by a historiography which has rarely engaged with the material premises of Eurasian growth. Recognizing the role of this forgotten industrial fuel means accounting for the statistical margins. The history of Russia's fossil economy is much more a history of regional diversity than the resource geographies of the present may want us to believe.

PART II

WORKING ENVIRONMENTS: EXTRACTION AND THE MAKING OF PLACE

3

Exploitation

From the point of view of Soviet bureaucracy, Sofia Polshkova was simply unlucky. In July 1952, the 22-year-old woman was working in a field where peat was spread out to dry when she was struck by lightning and passed away. The tragic incident happened while Sofia was working for the Shatura Peat Trust, whose extensive extraction sites stretched through the Meshchera Lowland in the south-eastern parts of the Moscow region bordering the neighboring region of Ryazan. At the time of the incident, the peat from Meshchera's wetlands had supplied the famous Shatura Power Station for nearly three decades. Glavtorf, the central peat administration within the Ministry of Electric Power Stations, did not attribute any particular significance to the woman's death. The staff member who received the telegram reporting the incident was quick to note that death from a lightning strike did not formally count as a professional accident (*proizvodstvennyi travmatizm*). The boundary between humans and nature, which this bureaucratic act implied, did not exist for Sofia Polshkova when she was doing her work. It mattered, however, for how her case was treated after she passed away. Polshkova's family was not entitled to any compensation. In official accident statistics, her sudden death would not even take the form of a number.[1]

[1] "Materialy rassledovaniia tiazhelykh neschastnykh sluchaev so smertel'nym iskhodom (1951–1953)," RGAĖ f. 7979, op. 2, d. 649, l. 166. On the reporting of industrial accidents and the pressure to keep them out of the official statistics, see Donald Filtzer, *Soviet Workers and Late Stalinism: Labour and the Restoration of the Stalinist System after World War II* (Cambridge: Cambridge University Press, 2002), 206–209.

For Tatyana and Nina Kazanova, employment in the peat industry ended in equally tragic circumstances. Picking up pieces of wood from the bottom of an open peat pit, the two women were helplessly exposed when frozen ground suddenly broke away from the upper edge of the pit and buried them underneath. It took their colleagues one and a half hours to uncover the women's bodies. Both were pronounced dead at the site. A subsequent investigation established that the technician Konovalova had reported the potential danger of the site to a higher-ranking employee a few days before her two colleagues died. But, against better knowledge, plans to blow up or remove the frozen ground to prevent it from collapsing once temperatures increased had not been carried out. Tatyana and Nina Kazanova lost their lives working in a spot that had already been deemed dangerous, their deaths being imbricated in a system of social relations in which their safety needs ranked below the fulfilment of production quotas. The incident happened in the Yaroslavl Peat Trust in May 1953.[2]

Workers like Sofia, Tatyana, and Nina were vital for making the peat riches of central Russia's wetlands consumable for the expanding fossil economy. From the late nineteenth century, ever-growing numbers of people became involved in the highly labor-intensive preparation of peat fuel, which served as a source of heat and, since the 1920s, electric power. Until the late 1950s, most of them were seasonal workers from the countryside who came to the extraction sites for a couple of months every year. In 1950, according to estimations based on Soviet records, some 215,000 people labored in peat extraction and ancillary activities, often with nothing but their arms, backs, legs, and shoulders to complete their work.[3] Employment rates began to decline soon after that, first because machines started taking over more and more tasks previously carried out by hand and, toward the end of the Soviet period, because peat-fueled power utilities transitioned to natural gas arriving through newly built pipelines from Siberia. The new energy geography emerging from this shift was one marked by the long-distance travel of fossil fuels from the places of extraction to the places of combustion. Peat and the economic regionalism that had been embedded in the Soviet energy system since the 1920s would consequently lose in significance.

[2] "Materialy rassledovaniia neschastnykh sluchaev s tiazhelym iskhodom (1953)," RGAĖ f. 7969, op. 2, d. 654, ll. 176–200.

[3] Demitri B. Shimkin, *The Soviet Mineral-Fuels Industries, 1928–1958: A Statistical Survey* (Washington, DC: US Department of Commerce, 1962), 87.

Peat extraction was certainly not among the deadliest industries in the Soviet Union. Accidents during which one or several workers lost their lives did not happen frequently and were usually attributed to individual neglect or inevitable force majeure events. The unspectacular, quiet deaths of Sofia Polshkova and Tatyana and Nina Kazanova point to the unequal social and gendered distribution of risk at their workplaces – and within society more widely. The historian Thomas G. Andrews coined the term "workscapes" to grasp the tangible and intangible elements constituting the working environments of coal miners in the United States. Emphasizing humans' inability to fully dominate their physical environment and the role of material factors in social and political histories of labor, this concept is pertinent to the forgotten history of peat in Russia's fossil economy.[4] The three female peat workers who tragically lost their lives had nothing to protect themselves from the danger rooted in the unstable relationships between human bodies, technology, and wetland environments, which formed and constantly evolved in peat extraction sites. These unruly workscapes exposed the women to forces of nature completely independent of human control, such as lightning strikes, and to the agency of matter unleashed by the imperative of extraction, such as frozen earth collapsing from the pit's edge (Figure 3.1). But the women's vulnerability had causes that lay beyond the concrete material circumstances under which they lost their lives. Ultimately, the nature of their deaths was social, a reflection of peat workers' marginal position within the economic fabric of imperial and Soviet Russia, where the countryside functioned as a reservoir of cheap labor and cheap resources for the overriding objective of economic expansion.

This chapter emphasizes the role of a largely forgotten kind of extractive labor in the history of industrial development and electrification in central Russia. Following peat workers on their way to and through their everyday at work, I explore the extraction of peat as a social practice and a more-than-human experience at the same time. Russia's fossil economy, with its ever-growing demand for fuel, relied heavily on the exploitation of human labor as a source of energy while also being shaped by the physical geography and environmental processes external to the human world. Emerging from the interplay of these factors was what the

[4] Thomas G. Andrews, *Killing for Coal: America's Deadliest Labor War* (Cambridge, MA: Harvard University Press, 2008), 125. On disasters in the hazardous environments of coal mines, see most recently Victor Seow, *Carbon Technocracy: Energy Regimes in Modern East Asia* (Chicago: The University of Chicago Press, 2021), 108–112.

FIGURE 3.1 Drying and stacking peat, 1950s. The place where Sofia Polshkova died in 1952 probably looked very similar. *Source:* Russian State Film and Photo Archive (RGAKFD). Reproduced with permission.

anthropologist Anna Tsing calls a "polyphonic assemblage" constituted by multiple intersecting actors – human and nonhuman – and shaped by multiple overlapping temporalities that upend linear narratives of progress.[5] Nonhuman rhythms of time and nonhuman agencies permeated all stages of the process during which peat turned into fuel, and they reverberated in the lived experiences of the workers who exploited the carbon stocks of central Russia's wetlands. In the peat industry, these factors co-shaped the social relations of extraction, marked for nearly a

[5] Anna Lowenhaupt Tsing, *The Mushroom at the End of the World: On the Possibility of Life in Capitalist Ruins* (Princeton: Princeton University Press, 2015), 23–25.

century by the prevalence of seasonal labor drawn from rural regions, and they challenged the vision of linear industrial time, particularly when the weather impeded or slowed drainage or extraction works. The peat fuel, which allowed regional power stations and machines in industrial enterprises to operate, embodied the unequal relationships between rural regions and places of dynamic industrial development. At the same time, it was the product of the complex interactions between people, technology, and matter in the more-than-human workscapes of the peat industry. It is by attending to the messy, often unruly constellations between the social and material worlds and the experiences of those confronting these in their daily lives that we see how dependent Russia's fossil economy was on the simultaneous exploitation of people and land at its margins.

RHYTHMS OF NATURE IN AN INDUSTRIALIZING ECONOMY

Peatlands do not readily give up their peat. In fact, peat can be turned into fuel only after a series of intrusive activities through which humans fundamentally alter the characteristics of a peatland: the digging of canals to drain the land, the clearance of the surface vegetation followed by the actual excavation of peat and, finally, the drying and stacking of peat before it will be stored or shipped to its various end users. In central Russia, where winters used to be cold and summers short, all these tasks had their season. Most work fell into the warm and dry months between early May and late July. The drying and stacking of peat usually took until September. The timing of these tasks was not entirely predictable though. A long winter, when the ground was frozen well into spring, could delay the beginning of the extraction period. Rainy weather, in turn, could extend peat works deep into the fall to allow the moisture in the peat to evaporate. Seasonality, with all its irregularities, profoundly shaped the experiences of workers in the peat pits and managers who tried to square the unruly rhythms of nature with the energy demands of an industrializing economy. Seasonality survived as a defining trait of the peat industry as the country underwent massive social, economic, and political change. From the late nineteenth century until the post-Stalinist period, the rhythms of nature reverberated in the rhythms of work at the margins of Russia's fossil economy.

Most of the seasonal workers came from the village. In the late imperial period, hundreds of thousands of peasants left their homes each year to take up temporary employment in urban factories or proto-industrial enterprises in the countryside. Seasonal labor migration (*otkhodnichestvo*) allowed them to pay redemption fees for the land they had received following the end of serfdom and to buy additional land or manufactured goods.[6] Peat extraction sites were an important destination for peasant workers in late imperial Russia. Gathering together in temporary associations, known as artels (Russ. sing. *artel'*), women and men from the villages usually spent the weeks between the sowing and the harvest periods extracting peat, their wage tied to the volume of peat prepared by the artel during a season.[7] The contemporary economist Lev B. Kafengauz estimated that, in 1888, there were just under 15,000 peat workers in Russia but, by 1913, their number had increased to 56,000. In comparison, just under 45,000 people were involved in oil drilling in Russia's thriving petroleum industry in the same year.[8] The presence of peat workers was particularly prominent in the extensive wetlands along the Nizhny Novgorod Railways, which operated between Moscow, Vladimir, and Nizhny Novgorod. In the 1890s, thousands of workers produced peat fuel for the booming industries in the region, most importantly textile mills.[9] Seasonal peat extraction was so widespread in those years that the future revolutionary leader Vladimir Lenin mentioned it in his *The Development of Capitalism in Russia* as an example of the "appendage to the factory," a form of nonindustrial wage labor driven by the demands of industrial production.[10]

These cycles of mobility and extraction survived the collapse of the Tsarist regime and the Bolshevik takeover in 1917. While the Soviet leaders pursued a radical overhaul of the social, economic, and political order, little changed for those who migrated between farming and peat extraction. In fact, the link between these two spheres became even tighter as more people from the countryside got involved in the mobilization of

[6] Gijs Kessler, "The Rural–Urban Nexus in Russian Labour History, 1860s–1930s: Suggestions for a Global, Comparative History," in Marcel van der Linden and Prabhu P. Mohapatra, eds., *Labour Matters: Towards Global Histories. Studies in Honour of Sabyasachi Bhattacharya* (New Delhi: Tulika Books, 2009), 207–225.

[7] Joseph Bradley, *Muzhik and Muscovite: Urbanization in Late Imperial Russia* (Berkeley: University of California Press, 1985), 28–29; Jeffrey Burds, *Peasant Dreams and Market Politics: Labor Migration and the Russian Village, 1861–1905* (Pittsburgh: University of Pittsburgh Press, 1998), 110, 122–123.

[8] Kafengauz, *Ėvoliutsiia promyshlennogo proizvodstva*, 337.

[9] V. Karsakov, "Chem otaplivat' parovozy? Zametka o vozmozhnom razvitii torfianogo dela v sviazi s nuzhdami gosudarstvennogo i zheleznodorozhnogo khoziaistva," in P. M. Solov'ev, ed., *Torfianoe delo: Obzor vydaiushchikhsia novostei po kul'ture bolot i torfodobyvaniiu za poslednee vremia* (Saint Petersburg: Tipografiia Berman i K°, 1897), 68–79.

[10] V. I. Lenin, *Collected Works Vol. 3: The Development of Capitalism in Russia* (Moscow: Progress Publishers, 4th printing 1977), 534.

peat for industry and power generation in the early Soviet period. Seasonal labor migration, which had declined dramatically during the First World War and the revolutionary period, bounced back during the New Economic Policy (NEP). In the late 1920s, some 127,000 migrant workers moved back and forth between their villages and peat extraction sites in the Soviet Union every year.[11] Enterprises actively attracted workers, sending recruiters to the villages late in autumn or during the winter, where they hired groups of peasants for the next season. In some areas, recruiters could draw on already long-existing collectives of seasonal migrants. When Soviet writer Mikhail Prishvin visited a peat extraction site in the region of Yaroslavl during the mid-1920s, he encountered an artel whose members had worked and travelled together for more than 20 years.[12]

Stalin's drive for collectivization and industrialization in the early 1930s, a watershed moment in the history of the modern Russian economy, did not sever the ties between farming and peat extraction either. Responding to the changing social and political realities, peat companies now concluded contracts with the newly created collective farms to extract labor from the countryside. In 1935, roughly 90 per cent of Shaturtorf's seasonal workers came from collective farms.[13] Hiring workers, however, became increasingly difficult in the years immediately following Stalin's collectivization campaign. As millions of peasants left the village to escape violence, hunger, and turmoil and to take advantage of new employment opportunities in construction or industry, peat companies encountered fierce competition for women and men able to work. Meeting recruitment targets was also challenging because Kolkhoz directors, under pressure to fulfil their production

[11] Gijs Kessler, "The Peasant and the Town: Rural–Urban Migration in the Soviet Union, 1929–40," PhD Dissertation, European University Institute (2001), Appendix B, Table B-1. On seasonal migration in the early Soviet period and the state's efforts to control it, see Lewis H. Siegelbaum and Leslie Page Moch, *Broad Is My Native Land: Repertoires and Regimes of Migration in Russia's Twentieth Century* (Ithaca, NY: Cornell University Press, 2014), 75–85.

[12] M. M. Prishvin, "Torf," in *Sobranie sochenenii, Vol. 3, Proizvedeniia 1924–1935 godov* (Moscow: Khudozhestvennaia Literatura, 1983), 497.

[13] Shatura Peat Trust, Annual report (1935), RGAĖ f. 7969, op. 1, d. 123, l. 191. For a case study on the recruitment and mobility of Mordovian peat workers in the Stalinist period, see N. A. Rodionov, "Voprosy ispol'zovaniia trudovykh resursov Mordovskoi ASSR v torfodobyvaiushchei promyshlennosti Udmurtskoi Respubliki (1940–1960-e gody)," *Ėkonomicheskaia Istoriia* 17, no. 2 (2012): 53–64.

quota, were keen to keep as many people on their farms as possible instead of sending them away.[14] Recruiters now often travelled hundreds, sometimes thousands of kilometers to mobilize workers for the peat season. As a result, very little was planned or coordinated in the recruitment of labor and the production of peat fuel during the first Five-Year Plans. Typically, fewer seasonal workers arrived at extraction sites than specified in the contracts between peat companies and collective farms. Of the 21,324 men and women contracted in the regions of Moscow, Ivanovo and Tatarstan, just 17,490 took up their work in the Shatura Peat Trust in 1935. The others had either stayed on their farms or taken up work elsewhere.[15]

In this complicated web of social hierarchies and dependencies, seasonal workers frequently deployed the "weapons of the weak" to leverage their interests. In 1923, around 13,000 peat workers went on strike in the Moscow region alone. Peat works in Tver and Ivanovo-Voznesensk saw labor protests in the same season, too.[16] Workers' dissatisfaction did not evaporate over the years. Throughout the 1920s and the early 1930s, the secret police recorded multiple instances in which seasonal peat workers destroyed machinery, went on strike, or quit in protest against low pay, insufficient food provisions, and high labor norms. Entire artels left their workplace for good when their demands were left unsatisfied.[17] Workers' early departures, whether in protest or to travel back to their villages for the harvest season, revealed the notion of peat extraction as a smoothly running industry to be mere administrative fiction. In July 1931, an article prominently placed on the first page of the major Soviet newspaper *Pravda* called for peat workers to return home only after all targets were met and the season officially declared over. The appeal emphasized some recent policy changes introduced by the Soviet leadership to make seasonal migrants more invested in their work. *Otkhodniks* would not have

[14] On labour recruitment in the countryside following collectivisation, see Sheila Fitzpatrick, *Stalin's Peasants: Resistance and Survival in the Russian Village after Collectivisation* (New York: Oxford University Press, 1994), 164–173.

[15] Shatura Peat Trust, Annual report (1935), RGAĖ f. 7969, op. 1, d. 123, ll. 189–192.

[16] Simon Pirani, *The Russian Revolution in Retreat, 1920–24: Soviet Workers and the New Communist Elite* (London: Routledge, 2008), 157, 201.

[17] These instances are documented in *"Sovershenno sekretno": Liubianka – Stalinu o polozhenii v strane (1922–1934 gg.)*, 10 volumes (Moscow: IRI RAN, 2001–2017), Vol. 4, part 1, 308, 310, 331; Vol. 5, 357, 359, 384; Vol. 6, 381; Vol. 7, 370; Vol. 8, part 1, 254, 311–314, 338, 374–376, 446, 505; Vol. 8, part 2, 1044–1045; Vol. 10, part 2, 636–637.

to surrender their salaries to the Kolkhoz farm, while the tax rate on off-farm income would be reduced.[18]

Since unauthorized departures remained the norm rather than the exception, the state's efforts to align the rhythms of peat extraction with the demands of an energy-hungry economy grew increasingly draconian – a trend that permeated all parts of the Soviet economy. During the 1930s, at least 35 per cent of Shaturtorf's seasonal workers would quit before their contracts ended. In 1940, following a notorious decree that criminalized workers who stayed absent from or left the workplace without permission, the trust reported that "labor discipline" was improving.[19] Recruiters, whose tasks became much more challenging as millions joined the Red Army during the Second World War, also felt the state's coercive approach to controlling Soviet workers. Shaturtorf's representatives now had to venture ever further afield to contract men and women for the upcoming peat season. For many, breaking the law became almost inevitable. In 1942, the employee Shcherbakov came back from the region of Voronezh, where he hired 69 instead of 330 workers for Shaturtorf's enterprise Petrovsko-Kobelevskoe. His colleague Gnezdova, also recruiting in Voronezh, made contracts only with 84 workers out of a planned 130. Egorov traveled more than 1,000 kilometers north to Arkhangelsk, where he secured 100 workers for the Shatura Peat Trust, a number significantly lower than his target of 365. Even worse, they all returned prematurely and without official permission. In strict adherence to the new labor code, the Central Peat Administration in the People's Commissariat of Power Stations mandated that all of them be fired and prosecuted, a frightening prospect they shared with recruiters from other peat trusts.[20]

Despite the state's tightening grip on their work and migration patterns, spending a few months each year working in peat extraction was an attractive prospect for men and women in the countryside, a rare chance to alleviate some of their daily hardship. Interviewed for the Harvard

18 "Kazhdyi torforabochii dolzhen znat' dekret o l'gotakh otkhodnikam," *Pravda* (3 July 1931): 1. See also Siegelbaum and Moch, *Broad Is My Native Land*, 82.

19 Shatura Peat Trust, Annual Report (1940), RGAĖ f. 7969, op. 1, d. 297, l. 260. High levels of labour fluctuation were very common in the early Stalinist period and, indeed, remained a persistent feature of the Soviet economy for decades. Alec Nove, *An Economic History of the USSR, 1917–1991*, 3rd ed. (London: Penguin, 1992), 199–201. On the criminalisation of workers in 1940, see ibid., 264–267.

20 "Prikaz po Glavnomu Upravleniiu Torfianoi Promyshlennosti NKĖS SSSR (1942)," RGAĖ f. 7969, op. 2, d. 5, l. 80ob.

Project on the Soviet Social System in 1950, a displaced woman from the Belarussian Soviet Republic recounted that being chosen for outside work in the peat industry was a matter of favoritism on her collective farm:

> there was a peat factory that needed workers only during the summer, and the Kolkhoz had to send a certain number of workers there. So each summer the Kolkhoz chairman would select people he knew best or people he liked, or relatives, and would let them go to work in the factory.[21]

This was a common experience. Kolkhoz workers' relationship with their chairman often decided whether or not they would be granted permission to engage in outside work. As a result, Kolkhoz workers who got the opportunity to do so frequently attracted envy among those who stayed in the village throughout the year without access to outside earnings and goods.[22]

The benefits of temporary employment in peat extraction were real. When companies fulfilled their side of the contract, peat workers earned rubles and goods, a clear contrast to the intangible "labor days" they received on their collective farms. In the autumn of 1934, Shaturtorf's weekly newspaper *Za Torf* reported on the departure of the last seasonal workers preparing for their journey home. Forewoman Kostiushina wrote excitedly about the women in her brigade (the term had replaced the prerevolutionary artel) who all earned a salary which, depending on their performance, ranged between 146 and 245 rubles per month. Money was not the only thing the women received. They each carried 15 meters of fabric, dried fish, sugar, a pair of boots, and galoshes. Yet, what Kostiushina praised as the workers' triumphal return to their farms was, in fact, a transfer of some of the most basic goods to allow people in the village to feed, wash, and dress themselves. Even two bars of soap per worker were worth reporting, as were the 5.2 kilograms of plain flour for those workers whose performance had exceeded the norm by 40 percent.[23] While seasonal *otkhod* had always helped to improve life in the countryside, Stalin's assault on the peasant economy created

[21] Harvard Project on the Soviet Social System. Schedule A, Vol. 8, Case 108, page 23, seq. 24, available at https://iiif.lib.harvard.edu/manifests/view/drs:5210929$24i.

[22] On favouritism and conflict surrounding *otkhod* in the Stalinist period, see Fitzpatrick, *Stalin's Peasants,* 164–173 and Jean Lévesque, "'Into the Grey Zone': Sham Peasants and the Limits of the Kolkhoz Order in the Post-War Russian Village, 1945–1953," in Juliane Fürst, ed., *Late Stalinist Russia: Society between Reconstruction and Reinvention* (London: Routledge, 2006), 103–119, 115.

[23] "Vesel'e i radost', bodrost' i ėnergiiu," *Za Torf* (October 23, 1934), 4.

conditions under which such additional income turned into a matter of survival.

Peat companies could also be a springboard for peasants who tried to leave behind their lives in the traumatized and poverty-ridden villages. This happened, for instance, when someone found seasonal work in peat extraction and then moved on to a year-round industrial workplace, a path taken by many peat workers in the region of Moscow during the turmoil of the early 1930s.[24] Another option was to take on a permanent position in a peat trust, where the cold months were used to serve and repair machinery. The transcript of an interview with another Soviet displaced person mentions a case from Belarus, where a farmer who had resisted the pressure to join the collective farm ultimately decided to leave the village for good:

> … in 1936 the government took away from him the cow and the horse. Now he was finished. After three days of that shameful dekulakization he took his children and wife and left to dig peat. During one day he dug up 1500 pieces of turf. For his work he received 5–6 rubles a day. But it was awfully hard work. The only advantage of that work was the possibility to buy a bowl of soup and some bread.[25]

Tracing the further trajectory of this man and his family is impossible, but the phenomenon seems to have been common enough as it features in one of the most famous literary representations of the Stalinist period. Gulag inmate Ivan Shukhov, the protagonist in Alexander Solzhenitsyn's *One Day in the Life of Ivan Denisovich*, learned through a letter from his wife that peat companies were a common destination among those who wanted to turn their back on the countryside: "not a single living soul had joined the *kolkhoz* since the war: all the young lads and girls had somehow wangled their way to town to work in a factory, or else to the peat workings."[26]

When the Soviet Union finally emerged from the destruction and trauma of famine, terror, and war, the principal parameters defining the exploitation of central Russia's peatlands and their place within the

[24] David L. Hoffmann, *Peasant Metropolis: Social Identities in Moscow, 1929–1941* (Ithaca, NY: Cornell University Press, 1994), 69.

[25] Harvard Project on the Soviet Social System. Schedule A, Vol. 34, Case 380/(NY)1460, page 34, seq. 34, available at https://iiif.lib.harvard.edu/manifests/view/drs:5605044$34i. Accessed March 24, 2025.

[26] Aleksandr Solzhenitsyn, *One Day in the Life of Ivan Denisovich*. Translated from the Russian by H. T. Willetts. Foreword by Alexis Klimoff (London: Vintage Books, 2003), 33.

country's energy system were still in place. At the time of Stalin's death in 1953, agriculture and industry were much less entangled than during the early phases of Russian industrialization. But peat extraction, very much like forestry, which relied strongly on seasonal workers from a rural background, kept blurring the boundaries between "traditional" and "modern" modes of production and the temporal regimes (cyclical vs. linear) conventionally ascribed to them.[27] Given that large-scale peat extraction could take place only when the soil was not frozen and that many tasks still required manual labor, Soviet peat trusts remained tied to the rhythms of the seasons as well as to the mobilities of short-term workers who moved back and forth between farming and extraction every year. The endurance of these links even showed at the level of workers' individual experience. In 1957, Shaturtorf's weekly newspaper reported on a brigade of women who had come to the same enterprise over a period of 15 years.[28]

For the workers, supplementing their meagre household budgets remained the key motivation to extract peat during the summer. In the post-war decades, the circumstances differed from those of the early Stalinist period in that goods of primary necessity gradually lost their dominant role in these transfers. Still, outside earnings remained highly attractive, particularly to peat workers from collective farms, where, other than in state farms, regular monetary payments were only introduced in the mid-1960s.[29] Seasonal labor migration allowed collective farm workers to channel money to their families or to convert their salary directly into consumer goods. In the late 1950s, cotton and silk fabric, woolen dresses, modern jackets, radios, watches, and bicycles traveled through Shaturtorf's enterprises.[30] One can hardly overstate the significance of these things, which could make one's life in the countryside more convenient or serve as a medium of barter exchange. In a series of oral histories recorded by a local museum of the El'niki district in Mordovia,

[27] Kateryna Burkush, "On the Forest Front: Labour Relations and Seasonal Migration in 1960s–80s," *Labor History* 59, no. 3 (2018): 295–315; Kochetkova, *Green Power of Socialism*, 14–15.

[28] "Torfianiki zakliuchaiut trudovye dogovory o rabote na torfopredpriiatiiakh v sezone 1958 g.," *Za Torf* (August 8, 1957), 2; "15 let na torfu," *Za Torf* (August 29, 1957), 2; "750 torfianikov i torfianits zakliuchili trudovye dogovory," *Za Torf* (September 3, 1957), 2.

[29] On money in the rural economy after Stalin, see Alexandra Oberländer, "Hatching Money: The Political Economy of Eggs in the 1960s," *Cahiers du Monde russe* 61, no. 1–2 (2020): 231–256.

[30] "Promtovary – torfianikam," *Za Torf* (October 3, 1957), 1.

former peat workers vividly recall how earnings from peat extraction improved their daily lives. During the post-war years, many young women from El'niki went to extract peat in Nizhny Novgorod, Shatura, and even as far as the Belarusian Republic. Their return usually caused excitement, as it meant the arrival of fabric, kitchen tools made from aluminum, or food items in the deprived material world of the post-war village. In some cases, wages from peat extraction contributed quite substantially to the budget of a rural household. Maria Malozemova, born in 1939, was a seasonal peat worker for several years in the 1950s. Maria's father was one of the millions who lost their lives during the war. Her mother worked for the Kolkhoz farm. The young woman's wage was crucial to alleviate the family's poverty: "For the work, they paid us money, which I sent to my mother. This is how, little by little, we built ourselves a house."[31] Invisible in the grand narratives of industrialization, the conversion of peat into fuel left deep traces not only in the landscapes of central Russia but also in the material lives and memories of people from places that were hundreds of kilometers away, the rhythms of their work an echo of the distant cycles of extraction.

SHOCK OF THE OLD: CHEAP LABOR, GENDER, AND TECHNOLOGY

Human labor looms large in the history of Russia's fossil economy. Even though contemporary sources, particularly from the Soviet period, evoked the image of peat extraction as an industrial undertaking and referred to it as the "peat industry" (*torfianaia promyshlennost'*), the production of peat fuel had very little in common with a neatly functioning industrial setup of labor, time, and technology. The historian David Edgerton has prominently argued that histories of the modern world tend to reproduce simplistic linear accounts of technological innovation, underestimating the significance of "old" technologies and the fundamental role of manual tasks in a vast array of productive and reproductive activities.[32] Mirroring the argument about what Edgerton called "the shock of the old," energy historians have highlighted the role

[31] Elena V. Nikishova, *El'nikovtsy*, Vol. 3 (Saransk: K. Shapkarin, 2019), 200–203, 665–667. The quote is on page 667. I am grateful to Elena Nikishova from the municipal museum in El'niki for sending me a copy of this publication.

[32] David Edgerton, *The Shock of the Old: Technology and Global History since 1900* (London: Profile Books, 2008, first published 2006).

of organic sources of energy, particularly human labor, in a history of "additions" marking the rise of energy-intensive fossil economies.[33] Peat extraction in central Russia supports these arguments. The removal of peat from its wetland origin and its conversion into a combustible fuel for factories and power plants involved multiple tasks during which human muscles were the primary energy source to lift and move things, a pattern that persisted into the first decade following the Second World War. The continuous, and until the 1970s accelerating, generation of peat fuel for Russia's expanding industrial metabolism must therefore be read, first and foremost, as a history of the intertwined exploitation of cheap labor and cheap nature, which critical scholars such as Raj Patel and Jason Moore identified as a prerequisite of modern economic growth across the globe.[34]

Technology was simple at first. Up until the middle of the twentieth century, sod peat production was the most widespread method employed in peat extraction. Introduced first in the late nineteenth century, this approach involved elevators that lifted the peat from pits and molding machines that gave the peat a brick-like shape. Operating mainly with the help of portable steam engines, these machines received a lot of attention in contemporary expert literature, where they signified the arrival of modernity. In contrast to the progress narrative prevailing in such publications, human bodies remained crucial to mastering the many steps it took to convert peat into an industrial fuel. Workers dug drainage canals and ripped out bushes or trees during their long days. Sometimes up to their waist in water, they used simple shovels to cut and load the peat on the elevators. Moving the peat bricks to surrounding drying fields and then turning and stacking them were also tasks predominantly performed by hand. Not surprisingly, labor was the most significant expense factor in peat extraction. According to a 1911 survey undertaken among the owners and managers of peat works in central Russia, wages accounted for over 50 percent of the running costs associated with the preparation of peat fuel.[35]

How crucial a role cheap labor played in the mobilization of peat, and hence for the availability of cheap energy in central Russia, became

33 Bonneuil and Fressoz, *Shock of the Anthropocene*, 100–104. For reflections on the history of labor in the age of coal, see Barak, *Powering Empire*, 8–11 and subsequent chapters on water, animals, and humans. An evocative description of human labour in coal mining is provided in Seow, *Carbon Technocracy*, 52–62, 94–104.

34 Patel and Moore, *History of the World*.

35 *Trudy kommissii po izucheniiu torfianogo dela*, 87.

apparent once the First World War upended the social parameters of extraction that had emerged since the late nineteenth century. Industrial managers realized in the spring of 1915 that mass conscription presented a blow to the routines in peat extraction with its strong reliance on seasonal workers from the countryside. As millions of peasants joined the imperial army, long-established artels fell apart. After an unsuccessful attempt to postpone peat workers' mobilization until the end of the 1915 peat season, industrialists started negotiations with the Tsarist government to address the looming labor shortage. Anxious about whether there would be sufficient hands to cut, staple, and carry the peat required to keep their businesses running, factory owners quickly united to have their say in distributing a new source of cheap labor: prisoners of war (POWs). Throughout the month of April, influential entrepreneurs, such as the directors of the Savva and Vikula Morozov cotton mills, hurried to develop a scheme that would allow them to replace their missing workforce with captives. In the summer of 1915, they recorded their first success. After most regular peat workers had left extraction sites to return to their villages, some 200 POWs, most of them Czechs who had fought in the Austro-Hungarian army, completed the peat season in the province of Moscow.[36]

Unfree laborers kept up much of the work in peat extraction during the war, as they did in many other parts of the imperial economy, where the 15 million men drafted for the Tsarist army left industrial mills and farmers in a frantic search for working bodies. Many of the more than 2 million Austro-Hungarian and German soldiers in captivity worked on farms, construction sites, or in mines across the Russian Empire, confronting harsh living conditions, poor wages, and physical punishment. Some have estimated that 25,000 POWs lost their lives during the construction of the Murmansk railways.[37] Meanwhile, the allocation of

[36] Letter from the Society for the Support and Improvement of the Manufacturing Industry (1915), RGIA f. 425, op. 1, d. 49, ll. 83–84; M. N. Korelin, "Trud voennoplennykh na torfianykh razrabotkakh," *Vestnik torfianogo dela* 2, no. 1 (1915): 80–84; M. N. Korelin, "Trud voennoplennykh na torfianykh razrabotkakh," *Vestnik torfianogo dela* 2, no. 3–4 (1915): 426–443; "K otpusku plennykh na torfianye raboty v Moskovskoi gubernii," *Vestnik torfianogo dela* 3, no. 1 (1916): 122–123.

[37] Alon Rachamimov, *POWs and the Great War: Captivity on the Eastern Front* (Oxford: Berg, 2002), 107–115; Gatrell, *Russia's First World War*, 114; Verena Moritz, "Die österreichisch-ungarischen Kriegsgefangenen in der russischen Wirtschaft (1914 bis Oktober 1917)," *zeitgeschichte* 25, no. 11–12 (1998): 380–389; Reinhard Nachtigall, *Rußland und seine österreichisch-ungarischen Kriegsgefangenen (1914–1918)* (Remshalden: Verlag Bernhard Albert Greiner, 2003), 168–171.

POWs turned into a matter of tension and contest, as parts of the Tsarist administration and various groups of producers competed to secure as much labor for their own purposes as possible. Compared to agriculture, coal mining, and strategically vital infrastructural works, peat extraction ranked low in the government's priorities for distributing labor. Throughout the war, representatives of industrial enterprises and several state agencies complained that the interests of peat extraction were insufficiently considered when it came to the allocation of POWs. Despite their complaints, war captives worked on many peat extraction sites. In 1917, 28 per cent of all peat workers in Russia were POWs.[38]

Industrial producers tried to turn the situation to their own advantage. Peasants, who had accounted for the bulk of peat workers in peace times, were usually eager to leave extraction sites early enough to be back in the village for the harvest season. In rainy years, this could mean that wet peat remained in the fields and could not be stored or used as fuel. Much labor was also needed to turn trees, shrubs, and bushes ripped from the ground before or during the extraction works into highly sought-after firewood, activities which usually took place outside of the core peat season. Confronting the disruption of the workflow brought by the war, those overseeing peat extraction appreciated captive labor as a means to make themselves less dependent on peasant mobilities and cushion the effects of poor weather. When the central administration enquired in the summer of 1916 whether peat-extracting companies needed POWs beyond the end of the season, many responded affirmatively. The managers of the Voznesensky cotton mill northeast of Moscow, for instance, hoped to keep their 348 POWs employed in peat extraction, as most seasonal workers had already traveled home. Getting the chance to use the POWs, who represented 40 percent of all peat workers in the company overall, for an additional couple of weeks promised to increase peat and firewood production. The directors of a weaving and dying factory in the south-east of Moscow calculated similarly. Thirty of their 91 peat workers were POWs. If at least 20 stayed until the second half of October, they could make more firewood and prepare the peatland for the next extractive cycle.[39]

38 Gerald H. Davis, "The Life of Prisoners of War in Russia, 1914–1921," in Samuel R. Williamson Jr., and Peter Pastor, eds., *Essays on World War 1: Origins and Prisoners of War* (New York: Brooklyn College Press, 1983), 163–196, 175.

39 "Kopii oprosnykh listov torfodobyvaiushchikh predpriiatii srednoi Rossii (1916)," RGIA f. 425, op. 1, d. 64, ll. 75–76, 78–79.

The 1917 Revolution brought a change in discourse but not in substance, at least for those doing the hard work. The Soviet leadership feared that its massive reliance on manual labor could potentially undermine the escalation of peat extraction across European Russia and the ambition to make peat a pillar of Russia's revolutionized energy system. In his address to the All-Russia Congress of Soviets in December 1920, Lenin proclaimed that, under Soviet power, peat extraction would have to be mechanized for economic and ethical reasons alike:

> It must not be forgotten that we possess vast deposits of peat, which we cannot utilize because we cannot send people to do such back-breaking work. The capitalist system could send people to work under such harsh conditions. In the capitalist state people were driven to work there by hunger, but in the socialist state we cannot consign people to such intolerable work, and nobody will go there voluntarily. The capitalist system did everything for the upper crust. It was not concerned with the lower classes.[40]

Lenin and early Soviet planners like Krzhizhanovskii, the main architect of the GOĖLRO-plan with its orientation toward economic regionalism and what was called "local fuel" at the time, put high hopes into hydropeat *(gidrotorf)* extraction. Invented by the engineers Robert Ė. Klasson and Vladimir D. Kirpichnikov shortly before the First World War, this technique revolved around the use of high-pressure water hoses, which were directed at the peat and dissolved it into a slurry (Figure 3.2). Pipes subsequently pumped the watery peat mass through drains that could stretch over several kilometers onto fields where the peat would slowly dry out. Klasson believed that hydropeat extraction was particularly promising for exploiting peatlands in Russia and Scandinavia, where the roots of trees and shrubs badly hampered digging machines.[41] Until the 1950s, hydropeat was a widely employed extraction method in Soviet peat trusts. Internationally, this method never took off, though following the Second World War it was shortly practiced in Finland and in Ireland.[42]

40 V. I. Lenin, *Collected Works Vol. 31: April–December 1920* (Moscow: Progress Publishers, 2nd printing 1974), 510.

41 R. Ė. Klasson, "Hydrotorf," *Zeitschrift des Vereines deutscher Ingenieure* 68, no. 23 (June 7, 1924): 601–605; R. Ė. Klasson, "Gidrotorf v sviazi s raionnymi ėlektricheskimi stantsiiami," *Ėlektrichestvo* no. 5–6 (1923): 257–266.

42 Lea-Elina Nikkilä, "Hydropeat: An Old Harvesting Method," in Riita Korhonen, Leila Korpela, and Sakari Sarkkolo, eds., *Finland–Fenland: Research and Sustainable Utilisation of Mires and Peat* (Maahenki: Finnish Peatland Society, 2008), 106–109; Feehan and O'Donovan, *Bogs of Ireland*, 135.

FIGURE 3.2 Hydropeat extraction in the Soviet Union, 1948. *Source:* Russian State Film and Photo Archive (RGAKFD). Reproduced with Permission.

The expectation that hydropeat extraction would make peat fuel production less laborious was mere fiction, however. As water was added, not released, during the process, the method considerably prolonged the time the peat needed to get sufficiently dry. This, in turn, increased the number of workers required to oversee the drying process, which was one of the most labor-intensive stages on the long path from peat to fuel. After a tractor with special caterpillars cut the half-dry peat slurry into bricks, workers had to toss around the bricks until their moisture content had declined enough so they could be used as fuel. Stacking the peat bricks also remained a manual task. And since the absolute amount of extracted peat increased with the introduction of the hydropeat method, stacking, in turn, also required more workers. All this was good news for Lenin, who advocated "mechanized" hydropeat extraction against the backdrop of the fuel crisis. Hydropeat, he stressed, "eliminates the need for skilled workers, since even unskilled workers can perform the work under this method."[43]

Crucially, most "unskilled" workers involved in what was often depicted as auxiliary work were female. Men and women occupied strikingly different positions within the material and social worlds of

[43] Lenin, *Collected Works* Vol. 31, 510.

central Russia's peat extraction sites, a pattern that had emerged in the late imperial period but would become much more pronounced with the massive expansion in peat extraction under Soviet power. Male peat workers (*torfianiki*) would carry out most preparatory tasks, such as draining the land, removing surface vegetation, and getting the peat out of the ground. Men were also disproportionally involved in the operation or the supervision of machinery. Female peat workers (*torfianitsy* or *torfushki* as they were called in the vernacular) took care of the drying of peat before they put the bricks into baskets, carried them, and arranged the dry peat in stacks. This gendered division of labor was tied into the rhythms of rural life. As a rule, male workers returned to their villages earlier to engage in harvest work, while women often stayed until September and even October, until the peat was dry. The organization of labor along gender lines meant that men and women had quite different experiences of their work as an embodied, more-than-human activity.[44]

The Soviet peat industry not only inherited the gendered distribution of labor, but entrenched it even further. Occupational segregation was initially grounded in cultural norms about what were considered male and female types of labor. Maintaining and deepening it, however, became a deliberate economic strategy to expand output and keep production costs down.[45] At its roots lay the massive wage difference reflecting men's and women's different responsibilities. At the first hydropeat extraction sites in the early 1920s, female peat workers, all employed in the drying fields, earned 22.50 rubles. In comparison, quarry workers (*kar'ershchiki*) and those spreading the peat mass (*razlival'shchiki*) received 40 rubles.[46]

[44] Catherine Nash, "Environmental History, Philosophy and Difference," *Journal of Historical Geography* 26, no. 1 (2000): 23–27. This gendered division of labor is visible in the Illustrated Map of European Russia from 1896 (Figure 2.1, p. 68), which shows women stacking peat in the background while men work close to the excavator.

[45] On unequal pay during the 1920s and 1930s, see Melanie Ilič, "'Equal Pay for Equal Work': Women's Wages in Soviet Russia," in Melanie Ilič, ed., *The Palgrave Handbook of Women and Gender in Twentieth-Century Russia and the Soviet Union* (London: Palgrave Macmillan, 2018), 101–115. Occupational segregation and unequal pay remained important structural features of the economy until the late Soviet period, as shown by Alastair McAuley, *Women's Work and Wages in the Soviet Union* (London: George Allen & Unwin, 1981). On cheapness as a quality of female labor actively exploited by Soviet managers, see Shoshana Keller, "The Puzzle of Manual Harvest in Uzbekistan: Economics, Status and Labour in the Khrushchev Era," *Central Asian Survey* 34, no. 3 (2015), 296–309, 301–302.

[46] "Proizvodstvennaia programma Gidrotorfa 1921–1922," RGAĖ f. 4372, op. 5, d. 11, ll. 1–3ob.

The fact that women's labor was less expensive than men's certainly appealed to bureaucratic and technical elites, strengthening arguments in favor of innovations that, although freeing up male workers, created additional labor needs in areas considered the domain of female workers. In 1926, engineer I. A. Brodskii published a lengthy article about what he called the "human (*liudskoi*) capital" in the peat industry, in which he made exactly this point. Once the hydropeat method became more prevalent, Brodskii predicted, the gender balance would tilt toward women. Yet even though more workers would be needed to dry, carry, and stack peat, the technology still promised efficiency gains substantial enough to justify the additional labor needs. Given that "muscle power" (*muskul'naia sila*) would remain the dominant source of energy in the peat industry for the foreseeable future, tapping into the "underused" reservoir of rural women's labor was an economic imperative, Brodskii argued. Not only were women easier to hire than men. They also required much less training and could quickly work at full capacity, an important criterion in light of the government's commitment to rapidly accelerate peat extraction.[47]

Brodskii sensed the parameters that determined the social and material organization of the Soviet peat industry very well. Women would provide the bulk of the cheap manual labor that facilitated the dramatic increase in peat extraction following the adoption of the GOĖLRO-plan and throughout the Stalinist period. In 1939, 9,052 out of 11,309 seasonal workers joining Shaturtorf were women.[48] The trust's recruitment strategy underscored how ingrained occupational segregation had become in the peat industry. Just as men dominated tractor work in Soviet agriculture while women farm workers were more likely to perform manual tasks – a pattern that persisted for decades – most female peat workers were assigned works of allegedly secondary importance that were mechanized much later than the "core" tasks carried out by men.[49] Leonid A. Dolzhanskii, a student doing practical work in Shatura during the 1930s, observed that the most labor-intensive work, the stacking of peat, was exclusively done by women. Dolzhanskii also noted that the trust had shifted the payment system from compensating the women based on the

[47] I. A. Brodskii, "O 'liudskom' kapitale torfianoi promyshlennosti," *Torfianoe delo* no. 12 (1926): 260–266.

[48] Shatura Peat Trust, Annual Report (1939), RGAĖ, f. 7969, op. 1, d. 269, l. 44.

[49] McAuley, *Women's Work*, 112–115; Liubov Denisova, *Rural Women in the Soviet Union and Post-Soviet Russia*. Edited and translated by Irina Mukhina (Routledge: Abingdon, 2010), 13–23.

area of land they cleared by picking up peat to rewarding them for the volume of peat arranged in orderly stacks.[50] His observations echoed the regime's growing reliance on piece rates to increase labor intensity.[51] In the context of the peat industry, the pressure of expanding production with limited resources meant that managers made efforts to ensure that female labor retained its cheapness.

The Second World War reinforced these dynamics. The importance of female workers grew even further, as thousands were conscripted to produce fuel. Schoolchildren had to join the ranks of peat workers as well.[52] A woman from the region of Nizhny Novgorod, where the peat-fueled Balakhna Power Station delivered electricity to many munition plants, remembered how she was sent to the peat works after she had just turned 18 years old: "Oh, how hard the work in hydro-peat extraction was. Very few mechanisms – only hands, female hands, moreover.... Sometimes you'd think: they'd better call me into the army. But no, from the peat works they didn't take anybody."[53] Valentina Bushueva, who was a young woman during the war, later recalled her hard labor during the Siege of Leningrad, when the peat from the immediate surroundings of the city was decisive in producing electricity:

> It meant a deep pit and two very powerful fire-pumps on its edge. There was an engine and a pipe, and this pipe carried liquid for many kilometres. At the bottom of the pit there were squares called cards. When the engine started working, it produced peat liquid, which was pumped into pipes and carried to those squares. As for me, I stood there alone in that pit, you understand, and with a stick I drove away pieces of floating wood, icicles, and this in the month of May, and I am standing there up to my waist in that liquid. I have canvas trousers on, a canvas jacket with a hood, a canvas cape on top, and also boots reaching up to my thighs. They had to pull me out of there. Just imagine, twelve hours in a pit. When they brought us some gruel, they had to pull me out. But then we were given five hundred grams of bread there.[54]

[50] "Otchet o studencheskoi praktike v Shature (1931)," RGAĖ f. 9458, op. 1, d. 336, l. 24ob.

[51] Filtzer, *Soviet Workers and late Stalinism*, 233.

[52] L. V. Kopenkina, "Torfianaia otrasl' v gody Velikoi Otechestvennoi Voiny (k 70-letiiu Velikoi Pobedy)," *Trudy Instorfa* 13, no. 66 (2016): 50–56; Mikhail D. Kerbikov, "Torfushki: Vzgliad na istoriiu zabytogo truda," in K. A. Pakhaliuk, ed., *Georgievskie chteniia: Sbornik trudov po voennoi istorii otechestva* (Moscow: Iauza, 2021), 523–536.

[53] "Torfushki. Rasskaz Mani," recorded by Anna G. Lokhova. Online resource: http://radilov.ru/krayrodnoy/1248-torfushki.html. Accessed May 22, 2024.

[54] Quoted in Cynthia Simmons and Nina Perlina, eds., *Writing the Siege of Leningrad: Women's Diaries, Memoirs and Documentary Prose* (Pittsburgh: University of Pittsburgh Press, 2002), 136–137.

Apart from mobilizing women and children, the Soviet regime relied on unfree labor to keep peat extraction going during the Second World War. In the Ural region, where the need for fuel increased due to the relocation of munition factories from territories close to the frontlines, peat-extracting companies met their labor demands in part by using prison inmates and prisoners of war.[55] POWs continued working in peat extraction in the immediate post-war years. In July 1945, the People's Commissariat of Power Stations was due to receive some 15,000 war prisoners to support the labor force during the peat season.[56] In 2019, when I visited the workers' settlement Shaturtorf near the Shatura power plant, a local urged my friend and me to go and see what he called the "German graveyard." Following the man's advice, we ventured into the pine tree forest where, a kilometer or so outside of the settlement, we found crosses and plates installed by the German War Graves Commission commemorating some 120 German and Hungarian POWs who lost their lives there.[57] This largely forgotten history of forced labor points, once again, to one of the chief prerequisites enabling fossil fuel-powered industrial production: the ability to mobilize human labor at low cost (Figure 3.3).

Following the war, women kept fulfilling most of the tasks that required manual labor, exposing their bodies to the challenging routines of heavy physical labor as they waded through peat slurry or arranged peat bricks in high stacks. Although the war-related absence of men in the Soviet countryside, where peat-extracting companies drew most of their workforce, certainly impacted labor recruitment, demographic reasons alone do not explain the predominance of female workers employed to do "auxiliary" labor. In 1947, Rostorf, the chief administration of the peat industry within the Ministry for Fuel, had a total of 78,337 permanent employees, of which 40,505 (or 52 percent) were women. Despite their high share among the overall staff, women accounted for only 1,451 (or 23 percent) of the 6,272 technical workers who possessed the skills to operate, service, or repair machinery.[58] Shaturtorf's report from the same

55 N. A. Rodionov, "Mobilizatsiia torfianykh resursov Urala na nuzhdy voennogo proizvodstva (1941–1945 gg.)," *Istoricheskie, filosofskie, politicheskie i iuridicheskie nauki, kul'turologiia i iskusstvovedenie. Voprosy teorii i praktiki* 1, no. 1 (2015): 168–172.

56 Document 36 in O. V. Khlevniuk, V. A. Kozlov, and S. V. Mironenko, eds., *Zakliuchennye na stroikakh kommunizma: Gulag i ob-ekty ėnergetiki v SSSR. Sobranie dokumentov i fotografii* (Moscow: ROSSPĖN, 2008), 96–97.

57 https://kriegsgraeberstaetten.volksbund.de/friedhof/schaturtorf. Accessed March 24, 2025.

58 "Chislennost' rabochikh i sluzhashchikh po polu i vozrastu na 1 marta 1946 g.," RGAĖ, f. 7969, op. 2, d. 64, ll. 31–34.

FIGURE 3.3 Memorial site for Prisoners of War near Shatura. *Source:* Author's photograph (2019).

year confirms this pattern. Although roughly half of the trust's more than 12,000 employees were female workers, just 17 percent of Shaturtorf's technical staff were women.[59] These numbers are reflective of a much broader pattern. While Soviet propaganda painted a rosy picture of the emancipated working woman who joined the industrial labor force, gender differences remained deeply embedded in the organization of the workscapes of the peat industry as in the Soviet workplace more generally.[60]

[59] "Otchet o chislennosti rabochikh i sluzhazhchikh" (1947), RGAĖ f. 7969, op. 2, d. 64, l. 44.

[60] Susanne Conze, "Women's Work and Emancipation in the Soviet Union, 1941–50," in Melanie Ilič, ed., *Women in the Stalin Era* (Basingstoke: Palgrave, 2001), 216–234. Little would change in the Khrushchev period either. See Donald Filtzer, "Women Workers in the Khrushchev Era," in Melanie Ilič, Susan E. Reid and Lynne Attwood, eds., *Women in the Khrushchev Era* (Basingstoke: Palgrave Macmillan, 2004), 29–51.

It was the large-scale introduction of milled peat (*frezerny torf*) which would eventually undermine the importance of seasonal labor and the strong gender segregation that had formed the bedrock of industrial peat extraction since the late imperial period. Invented in the late 1920s, but widely adopted only from the late 1950s, milled peat represented an entirely new approach to the extractive exploitation of wetland environments. Following a cycle of intensive large-scale draining, specially equipped tractors would mill the surface of a drained wetland and break the peat soil into small crumbs, thereby erasing any visible traits that peat fuel originates from the waterlogged environment of a living wetland. Milled peat was burned in power stations or turned into briquettes, which rural households used for heating and cooking. In the late Soviet period, milled peat was the most important product of the peat industry.[61]

Particularly where peat trusts combined this new technology with harvest engines, which gathered the peat crumbs from the ground, cheap labor stopped being a defining feature of the Soviet peat industry. Planners noted in 1967 that the new method had effectively supplanted other extraction techniques, making seasonal workers largely redundant.[62] Shaturtorf's annual reports confirm this trend. Overall employment in the peat trust declined from the second half of the 1960s, while peat fuel production increased until the end of the following decade.[63] For those on the ground, the diffusion of milled peat extraction marked a turning point that changed not just the workflow but the demography of the peat industry. As a former employee of the Shatura Peat Trust, who had grown up and worked in the peat industry in Tugolesskii Bor just outside of Shatura, recalled during my visit to the region in 2016, "with milled peat, the women were not needed anymore."[64] The long-term nexus between peat extraction and the countryside as a prime supplier of cheap female labor was finally broken.

[61] L. I. Dlugotskii, P. V. Dolinin, and N. N. Uspenskii: *Frezernyi torf (Po materialam komissii SOIUZTORF po frezernomu torfu)* (Moscow: Nauchno-issled. Torf. Institut, 1931); N. V. Ivashechkin, "Torfianaia promyshlennost' RSFSR," *Torfianaia promyshlennost'* no. 12 (1972), 4–5.

[62] "Doklad Glavtorfa o sostoianii tekhnicheskogo urovnia proizvodstva v torfianoi otrasli za 1967 god," GARF f. A-632, op. 2, d. 134, l. 3.

[63] In 1965, Shaturtorf extracted a total of 3,039,000 metric tons of peat with a total of 9,057 employees. Peat extraction reached 5,270,000 metric tons of peat in 1977, when the trust had just 7,816 employees. Shatura Peat Trust, Annual report (1965), GARF f. A-632, op. 2, d. 35, l. 21, l. 87; Shatura Peat Trust, Annual report (1977), GARF f. A-632, op. 9, d. 148, ll. 106, 152.

[64] Collective interview with the author, Tugolesskii Bor, July 7, 2016.

But machines did not make peat fuel production entirely independent of the rhythm of the seasons. Labor needs in the workscapes of the peat industry kept fluctuating over the course of the year. Many companies continued hiring short-time workers for the summer months when most of the work had to be done. Compared to previous decades, when seasonal workers dominated the workforce, their numbers became smaller and smaller. In 1962, just under a third of all 70,000 employees in the RSFSR's peat industry were seasonal workers. In the same year, Gosplan and the Council of Ministers calculated that less than 13,000 collective or state farm workers would have to be recruited for the 1963 peat season in the entire republic.[65] At the local level, the decline in seasonal employment was drastic. Shaturtorf, which had attracted more than 20,000 seasonal workers through Orgnabor in 1944, hired just 333 people through this channel in 1967. Another 1,800 workers temporarily joined the trust outside of formal agreements because they had family working for Shaturtorf or other companies had sent them.[66] A result of technological changes within the peat industry, the shift away from seasonal labor was also indicative of a broader socioeconomic trend in the postwar Soviet Union. Labor in the countryside became increasingly scarce as more and more people left the villages to live in the cities. The strategy of extracting cheap labor from the countryside had clearly reached its limits.[67]

Meanwhile, mechanization posed new challenges. Depicted as proof that the peat industry was on a path toward efficiency and high productivity, milled peat extraction did not upend the cyclical nature of the workflow. At the same time, labor remained a limiting factor in peat fuel production. The shift to milled peat meant that enterprises required a new type of worker: *mekhanizatory*, people qualified to operate and service machines. In the late Soviet period, such employees were increasingly difficult to find. Shortages of machine operators, whether permanent or seasonal, became a constant headache for administrators, as experienced workers knew what they had to offer and readily packed their things when they got an opportunity to leave the peat industry behind. In the late 1970s, voluntary resignations in Shaturtorf's enterprises reached

65 "Ob obespechenii torfianoi promyshlennosti rabochei siloi na sezon 1963 goda," GARF A-262, op. 8, d. 1904, ll. 117–118, 120.

66 Shatura Peat Trust, Annual Report (1944), RGAĖ f. 7969, op. 2, d. 262, l. 94; Shatura Peat Trust, Annual Report (1968), GARF f. A-632, op. 2, d. 265, l. 131.

67 On rural–urban migration in the late Soviet period, see Siegelbaum and Moch, *Broad Is My Native Land*, 123–146.

unprecedented levels.[68] Following a common practice at the time, peat trusts now concluded contracts with vocational schools and universities to ensure they would have enough temporary workers in the summer.[69] After labor from the village had become less abundant, students now provided the cheap seasonal labor, without which the peat industry would have never been able to produce as much fuel as it did.

EXTRACTIVE LABOR IN AN UNRULY WORLD

Wherever workers extracted and processed the peat that would fuel Russia's expanding industrial metabolism, they acted as part of a complex and fluid set of relationships formed by the interplay of their bodies, power structures, and the material and immaterial elements of the nonhuman world. Physical immediacy was at the core of their experience, as they prepared a wetland for extraction or lifted, transported, and stored peat. The weight and texture of peat wrote itself into workers' bodies, making them archives of their daily hardships. Workers, moreover, had only limited control over the environment of their workplace. Peat extraction had a more-than-human dimension at all times, but it was in specific moments that nonhuman agents made their presence felt and interfered with production targets and workflows. Rain, storms, flooding, electric currents, or technical failure revealed that the boundaries between the material and the social worlds which the designers of these workscapes attempted to draw were never fully manageable. From a bird's eye, the peat industry was on a path of expansion from the late nineteenth century until the 1970s. Risk and uncertainty, however, remained persistent elements in its daily workings, affecting people unequally depending on their status both within the broader social fabric and the hierarchies of their workplace.

The weather was one of those factors which regularly undermined extractivist ambitions. Plenty of rain impeded the drying process and reduced the energy density of the peat; hot summers increased the risk of

[68] Shatura Peat Trust, Annual report (1976), GARF f. A-632, op. 9, d. 97, l. 98; Shatura Peat Trust, Annual report (1977), GARF f. A-632, op. 9, d. 148, l. 108; Shatura Peat Trust, Annual Report (1978), GARF A-632, op. 9, d. 198, l. 125.

[69] "Prikazy po Rostorfu za 1976 god," GARF A-632, op. 9, d. 53, l. 156–159; "Prikazy po Rostorfu za 1979 god," GARF f. A-632, op. 9, d. 203, l. 81; "Prikazy Rostorfa za 1980 god," GARF, f. A-632, op. 9, d. 252, ll. 89–90; "Prikazy po Rostorfu za 1981 god," GARF f. A-632, op. 9, d. 298, ll. 78–81. The importance of student workers, particularly in construction and farming, increased considerably in the late Soviet period. See William Moskoff, *Labour and Leisure in the Soviet Union: The Conflict between Public and Private Decision-Making in a Planned Economy* (London: Macmillan, 1984), 44–48.

fires, while long or early winters shortened the period suitable for excavation and drying works. Like yields in the farming sector, annual extraction and fuel production fluctuated considerably. In 1914, for instance, peat extraction in the Russian Empire reached record levels because of unusually favorable weather.[70] The weather was so important a factor that peat trusts' extensive annual reports always commented on the meteorological conditions and how these had impacted their operations. Take the following examples from Shaturtorf's annual records. In 1943, a storm blew away 11,430 tons of milled peat in one of the trust's enterprises; in 1955, following a winter with heavy snow, repeated flooding in the spring delayed works at the beginning of the season; the workers of Shaturtorf drilled less deeply than intended in 1965 because of unstable weather; in 1969, two of the trust's enterprises missed their targets because of "extremely poor meteorological conditions"; in 1976, which was a wet year, cold temperatures set in much earlier than usual, shortening the overall peat season.[71] Directors of peat trusts may have used the weather to hide other reasons behind their failure to fulfil the plan. After all, the weather was not something they could be held accountable for. Still, not even the highest representatives of the Soviet state doubted that peat fuel production, even after most of its steps had been mechanized, remained contingent upon factors outside their control. In 1974, in response to challenging weather conditions during the season, the Council of Ministers of the RSFSR retrospectively changed the planning targets for the peat industry.[72] While there were multiple structural causes constraining effective economic planning in the Soviet Union, the plan also faced limitations whose origins lay beyond the human world.

Workers felt the more-than-human dimension of peat extraction in much more intimate ways, their bodies becoming testimonies to the fact that the imperatives of growth and progress were often realized without much concern for the health of those who contributed their labor to achieve these abstract goals.[73] In the early Soviet period, thousands of

70 Kafengauz, *Ėvoliutsiia promyshlennogo proizvodstva*, 178.

71 Shatura Peat Trust, Annual Reports, RGAĖ f. 7969, op. 2, d. 241, l. 79 (1943); RGAĖ, f. 7979, op. 2, d. 539, l. 18 (1955); GARF f. A-632, op. 2, d. 35, l. 74ob (1965); GARF f. A-632, op. 2, d. 324, l. 94 (1969); GARF f. A-632, op. 9, d. 97, l. 97 (1976).

72 "Reshenie Ministerstva toplivnoi promyshlennosti RSFSR po godovomu otchetu Respublikanskogo promyshlennogo ob-edineniia po dobyche torfa Rostorf za 1974 god," GARF f. A-632, op. 2, d. 585, ll. 1–5.

73 This is an important thread running through Kate Brown's scholarship. For a poignant example, see Brown, *Manual for Survival*, 81–84.

peat workers got infected with malaria as they were toiling to provide fuel for the nascent power industry. Standing water and the presence of a highly mobile population living and working close to mosquito breeding sites made these places hotspots of the disease in central Russia.[74] But workers' bodies also ailed in very mundane ways. In contrast to the silence of official reports, which barely mention the physical discomfort of manual labor, workers' cultural production gives us a vivid picture of the daily hardships emerging from the interactions of technology, human bodies, and wetlands at peat extraction sites. In the early 1920s, the ethnographer D. I. Iskrin visited the region of Bogorodsk, one of the first centers of commercial peat extraction in late imperial Russia and a major provider of peat fuel throughout the Soviet period. Iskrin recorded *chastushki,* short songs which played an important role in the oral culture of the peasantry and often captured the voices of women.[75] For the female workers who speak through these songs, working in peat extraction meant constant pain rooted in a never-ending routine of picking peat up from the ground, raising large baskets, hauling them to where the peat was stored, unpacking the baskets, touching and lifting sod after sod to arrange them in a stack:

It's hard, oh so hard to bend down	Трудно, трудно нагинаться
And to lift sods of turf oh so black;	За черной торфиной;
It's harder still to stand up	А труднее подыматься
With a heavy creel on your back!	По ходам с корзиной!
In the bog my life is free:	На болоте жить привольно:
My shoulders ache from the creel;	От корзинок плечи больно;
My midriff feels less strain, –	На середине повольней, –
But my lower back is worse again.	Пояснице побольней!
In our bog the girls are famed	На нашем на болоте
Throughout the land:	Заслужили девки честь:
They've blisters on their hands,	На руках у них мозоли,
Sitting on their a***s causes pain.	А на ж... больно сесть.[76]

[74] Chapter 5 explores the history of malaria in the late imperial and early Soviet peat industry at greater detail.

[75] Isabel A. Tirado, "The Village Voice: Women's Views of Themselves and Their World in Russian *Chastushki* of the 1920's," *The Carl Beck Papers in Russian & East European Studies*, no. 1008 (Pittsburgh: The Center for Russian & East European Studies, 1993).

[76] D. I. Iskrin, *Na torfianykh bolotakh: Bytovoi ocherk po pesniam, zapisannym v Bogorodskom uezde* (Bogorodsk: Bogorodskii Institut Kraevedeniia i UProsa, 1925), 7. I am grateful to Sarah Smyth for assisting me with the translation.

Time did not bring relief. Despite Lenin's promise that Soviet power would eliminate the exploitative labor regime of late imperial capitalism, working conditions showed no notable improvement following the 1917 Revolution. At least for those who performed manual labor, peat extraction remained an extremely physically demanding activity. Moreover, in an effective way to keep labor cheap, administrative elites did not pay much attention to workers' daily needs. During the 1930s, Shaturtorf's female workers often missed the most basic means to fulfil their tasks. It was not unusual that they carried the peat in their aprons because the trust did not hand out enough baskets. Many women did not even have simple bast shoes, as they were common in the countryside, and walked barefoot, which made them suffer injuries on their feet. Shortages of petroleum jelly meant that they could not protect the skin of their hands from painful cuts and scratches.[77]

While there was little administrative effort to alleviate these issues, the heavy toll of labor on peat workers' bodies became a concern once it threatened to harm reproductive health. During the 1930s, the Soviet state actively pursued pronatalist policies and promoted a cult of motherhood, building on an earlier discourse which treated women's bodies primarily as medical objects and emphasized biological differences between women and men.[78] In line with these developments, health experts at the State Institute of Labor Protection investigated the impact of lifting and carrying heavy weights on the physical health of female peat workers. They established that problems of the menstrual cycle (irregular cycles, increased pain, or heavy bleeding) affected peat workers much more frequently than women performing manual labor in other economic sectors. Although most were young and had not yet delivered a child, many peat workers also displayed symptoms of a uterus prolapse. To investigate and

[77] "Ne v karzinakh (sic!), a v fartukakh my perenosim torf," *Za Torf* (August 10, 1934), 1; "My trebuem k sebe vnimaniia," *Za Torf* (August 18, 1934), 2; "Berezhno otnosit'sia k korzine," *Za Torf* (September 30, 1934), 3; "O bytovom obsluzhivanii torfianits na Bol'shom Gidrotorfe," *Za Torf* (August 18, 1934), 4; "Prikaz po Glavnomu upravleniiu torfianoi promyshlennosti (April 10, 1942)," RGAĖ, f. 7969, op. 2, d. 5, ll. 64–65.

[78] David L. Hoffmann, "Mothers in the Motherland: Stalinist Pronatalism in Its Pan-European Context," *Journal of Social History* 34, no. 1 (2000): 35–54; Yulia Gradskova, "Soviet People with Female Bodies: Performing Beauty and Maternity in Soviet Russian in the mid 1930–1960s" (PhD dissertation, Stockholm University, 2007), 98–102; Frances L. Bernstein, "'The Dictatorship of Sex': Science, Glands, and the Medical Construction of Gender Difference in Revolutionary Russia," in David L. Hoffmann and Yanni Kotsonis, eds., *Russian Modernity: Politics, Knowledge, Practices* (New York: St Martin's Press, 2000), 138–160.

remediate the problem, the researchers conducted an intrusive laboratory experiment, during which a water-filled balloon was inserted into a woman's vagina and linked to a writing nib which recorded the movements of the uterus while the woman performed different activities. They concluded that replacing baskets with stretchers (*nosil'ki*) carried by two women at a time would help prevent the adverse health effects facing female peat workers. The experts did not conceal that the research was economically motivated. From their point of view, the negative impact of heavy labor on female reproductive health presented a management issue rather than being a reflection of structural gender inequality. Carrying a weight of 50 kg of peat per stretcher, they suggested, would be optimal to allow women to use their bodies most efficiently without compromising health or output.[79] The larger significance of these studies is hard to miss. While the regime was anxious about the population's fertility, the desire to keep women in the workforce without curtailing productivity eventually took precedence, rendering women's bodies, like the peatlands where they worked, sites of extraction.

Health risks also came in less obvious forms. Electricity regularly exposed the illusion of human control over nature, which was an important, although not undisputed, trope in Soviet cultural imagination.[80] Electric power had inspired widespread enthusiasm for peat in the early Soviet period. In the day-to-day of the peat industry, however, it repeatedly acted as a life-threatening force. A particularly tragic incident happened in the southeast of the region of Moscow in 1926, during which a large group of seasonal peat workers from the province of Kaluga lost their lives. Their artel of 28 men was pushing a locomotive along narrow-gauge railway tracks when the pipe of the locomotive touched the lower cable of an overhead power line. Fifteen workers who had been holding on to the metallic chassis of the engine were electrocuted and died instantly.[81] Electricity remained a source of danger over the coming decades. In 1970, V. I. Nikandrova, who worked for a peat company in the region of Briansk, died when she was switching on the lights of a machine.[82] Two

79 I. Okuneva and E. Shteinbakh, *Ratsionalizatsiia i ozdorovlenie truda torfianits* (Moscow: Gosudarstvennoe sotsial'no-ėkonomicheskoe izdatel'stvo, 1931), 7–19.
80 Klaus Gestwa, "Das Besitzergreifen von Natur und Gesellschaft im Stalinismus: Enthusiastischer Umgestaltungswille und katastrophischer Fortschritt," *Saeculum* 56, no. 1 (2005): 105–138.
81 "Pamiati pogibshikh na sluzhebnom postu," *Torfianoe delo* no. 5 (1926): 1–2.
82 "Prikazy Glavtorfa po proizvodstvennoi deiatel'nosti za 1970 god," GARF f. A-632, op. 2, 336, ll. 128.

years later, electrician A. N. Piragov lost his life while repairing a power line for the Vladimir Peat Trust.[83] Electrocution was obviously not unique to peat-extracting enterprises. Humans have developed means to direct the flow of electric power, but it may still cross the boundaries of manufactured containment designed to protect human lives. In the history of Russian peat extraction, electric hazards presented one of many factors that interfered with human aspirations to create working environments in which everything was under control.

Widely celebrated for their ability to increase extraction rates and make workers' lives easier, machines could turn into unpredictable and potentially harmful agents as well. Even in its most mechanized form, peat fuel production involved many steps during which manual workers labored near running engines. Broken or poorly serviced machinery, a common problem in the Soviet economy, proved to be a particular safety risk.[84] In February 1953, 20-year-old Mavzida Islamova was gathering up sod peat with a spade when she was struck by the falling grab of a crane which had been loading peat onto the wagons of a narrow-gauge railway train. Islamova passed away on the site.[85] In 1968, V. M. Korolev, one of Shaturtorf's unskilled workers, died after a loose part of a milling machine hit him in the face.[86] Internal investigations brought up several mundane factors to explain these tragic incidents. In Islamova's case, the crane driver's assistant had not appeared at work, while two of Islamova's colleagues had been working at some distance from where they could not see the danger. In the case of Korolev, the dirty windows in the milling machine's cabin obstructed the driver's sight. More importantly for the investigators, both Islamova and Korolev had put themselves at risk by working in the wrong place. Korolev had been just forty-one instead of the prescribed fifty meters away from the milling machine, which, to make matters worse, had also lacked any posters with relevant safety instructions. That malfunctioning machinery created mortal danger for the workers does not seem to have mattered to the investigators in either case. Following Korolev's death, Shaturtorf's administration at least increased the safety distance between manual workers and drilling machines to 100 meters, a silent recognition that the machinery designed

[83] "Prikazy Glavtorfa za 1972," GARF f. A-632, op. 2, d. 444, ll. 78–79.

[84] See Filtzer, *Soviet Workers and Late Stalinism*, 218–220.

[85] "Materialy rassledovaniia neschastnykh sluchaev s tiazhelym iskhodom (1953)," RGAĖ f. 7969, op. 2, d. 654, ll. 87–106.

[86] "O neschastnykh sluchaiach, proisshedshikh na predpriiatiiakh Shaturskogo gosudarstvennogo torfianogo tresta (1968)," GARF f. A-632, op. 2, d. 202, ll. 51–52.

to maximize production could turn into a hazard on its own, particularly when spare parts were scarce and regular service neglected.

During the multiple steps involved in the production of peat fuel, workers were constantly in motion. They had to move across extraction fields, between extraction and drying fields, and between drying fields, peat stacks, and narrow-gauge railways, which delivered the peat to storage facilities or end users. The set-up of labor relations and the infrastructure facilitating these movements favored the quick relocation of peat over workers' mobility needs. A 1938 safety poster reminded workers to use only sturdy bridges across drainage ditches. At the forefront of the poster, a woman loses all the peat from her basket as she crashes through a makeshift bridge. Her much more diligent coworker crosses the ditch at a secure place, delivering her peat safely and without losing any time (Figure 3.4). As with any official materials relating to professional accidents, several readings of this source are possible. The poster implies that the woman in the foreground acted without the necessary precaution as she lazily decided to shorten her way. However, another conclusion is possible, too. That workers fell when crossing

FIGURE 3.4 "Don't walk over faulty bridges." Workers' safety poster.
Source: NKTP, Moscow, 1938.

drainage ditches was a common enough problem. Student Leonid Dolzhanskii, for instance, noted during his internship in Shatura that bridges on peat extraction sites were in very poor condition.[87] In the summer of 1953, Shaturtorf's tractor driver Mikhail Geraskin lost his life after he crossed a ditch at the wrong place and fell in.[88] Soviet administrators explained such incidents by pointing out workers' carelessness. However, the evidence also presents a clue as to administrators' hierarchy of priorities. What ultimately counted was the volume of usable peat, while dealing with the practicalities of moving around and transporting peat was left to lower-ranking employees. Under pressure to fulfil their quota or wishing to reduce the time spent carrying weighty burdens or getting around, workers used shortcuts. The risk of harming themselves apparently outweighed the hardships or pressures they faced otherwise.

Other cases suggest a similar reading. Lacking alternatives to get around when they were tired or had to overcome long distances, workers used railway wagons designated for transporting peat, sleighs, or tractor-trailers as a means of transportation. Such practices allowed them to accommodate their needs in an environment designed primarily to maximize fuel production. In most cases, this behavior seems to have served them well. However, moments when something went wrong laid bare how the neglect of workers' transport needs exacerbated their exposure to risk in the more-than-human workscapes of the peat industry. Take, for instance, the female workers sent to remove tree stumps from one of the milled peat fields of the Yaroslavl Peat Trust in April 1953. Having completed their task, the women loaded the stumps on a sleigh tied to a tractor and sat on the sleigh. In a bend on their way back, the sleigh suddenly turned over. Two women suffered injuries. Their colleague, A. T. Kobina, was killed.[89] In line with the official view of the incident, this could be interpreted as one of many situations where workers did not understand or knowingly bypassed safety rules. Yet, it could also be inferred that moving things seems to have been far more critical for the administration than moving workers. Possibly exhausted after having spent long hours working in the snow, the women had no choice but to sit on the sleigh to make their return somewhat easier. Three months later,

87 "Otchet o studencheskoi praktike v Shature (1931)," RGAĖ f. 9458, op. 1, d. 336, l. 14.

88 "Otchety i perepiska o neshchastnykh sluchaiakh za 1953," RGAĖ, f. 7969, op. 2, d. 652, l. 139.

89 "Materialy rassledovaniia neshchastnykh sluchaev s tiazhelym iskhodom (1953)," RGAĖ f. 7969, op. 2, d. 654, ll. 164–169.

Yargostorf's electrician Genadii Ogorelov, born in 1935, died tragically when he jumped onto a moving train wagon loaded with peat. The trust's administration at least admitted that Ogorelov's death reflected a more significant structural problem and ordered that regular passenger train services be introduced to the trust's enterprises.[90]

Accidents with interns and student workers point to another professional hazard in the peat industry. In the late Soviet period, student workers became a common sight on peat extraction sites, where they worked alongside regular employees. Explicitly encouraged by central planning organizations, where the increasing difficulty of drawing cheap labor from the countryside was well understood, student employment could take a tragic turn when company directors deployed students to compensate for the chronic lack of machine drivers. During her internship at Lengostorf's enterprise Naziia in 1969, F. D. Margolina, a student at the Polytechnic Institute in Minsk, was hit by a moving tractor and pushed under a tedder usually used to turn over peat that was getting dry. The woman passed away from her injuries.[91] Intern Rzhanikova, like Margolina, a student sent to a peat company in the Leningrad region, died in very similar circumstances as she was struck by a tractor that was sliding down a hillside where the young woman had improperly parked the machine.[92] In 1968, the student Korotaevskaia, when practicing how to drive a tractor, ran over her fellow student worker Smurova, who died from her injuries.[93] None of these deadly incidents are specific to the peat industry. In fact, they could have equally happened in the farming sector or in construction, which were similarly dependent on temporary student labor in the late Soviet period. Still, they are telling examples of the economic logic that had dominated the peat industry from its beginnings. In all cases, internal investigations established that the students, who were given the same tasks as regular employees, had not received sufficient training. Cheap labor was so crucial a factor in the peat industry that managers were willing to ignore workers' needs and safety. Most times, this practice allowed them to muddle through, but it could lead to such tragic outcomes as the death of students who were unprepared to assess

90 "Rassledovanie neshchastnykh sluchaev s tiazhelym iskhodom (1953)," RGAĖ, f. 7969, op. 2, d. 655, ll. 9–10.

91 "Prikazy Glavtorfa proizvodstvennye za 1969 god," GARF f. A-632, op. 2, d. 277, ll. 50–60.

92 "Prikazy Glavtorfa za 1972," GARF f. A-632, op. 2, d. 444, ll. 80–81.

93 "Prikazy Glavtorfa za 1968 god," GARF f. A-632, op. 2, d. 202, ll. 28–30.

or face the risks in the more-than-human environments of peat extraction sites.

The examples of production failure, health issues, and deadly accidents discussed here may appear as random individual cases without any larger significance. Soviet administrators certainly seem to have taken such a view. They routinely blamed factors outside human control (the weather, natural disasters, or acts of god, *stikhiinye bedstviia*) or the workers themselves (their ignorance, imprudence, and lack of responsibility). Seen together, however, these episodes foreground the entanglements and repercussions between human bodies, socially constructed hierarchies of power, and nonhuman agencies, which conventional economic histories fail to account for. Within peat-extracting companies, the risks linked to the unpredictable interplay of these factors were distributed unevenly. Bad weather may have put managers in the uncomfortable situation of having to explain their failure to meet quotas to higher-ranking authorities, but workers paid with their physical well-being and sometimes even their lives for the exposure of their bodies to the presence of nonhuman agencies in their workday. Recognizing its more-than-human dimension, we may appreciate how unequal a social experience peat extraction and, hence, the production of fuel for Russia's expanding fossil economy was.

* * *

Industrial expansion and the mobilization of energy resources underpinning it looked different in different parts of the world. In central Russia, these processes hinged at least in part on the carbon stocks that had accumulated in Eurasian wetlands over thousands of years and on the cyclical mobility of tens and, at its peak, hundreds of thousands of workers who moved back and forth between food and fuel production every year. Inscribed in the history of Russia's fossil economy were experiences of inequality and marginality that have been largely forgotten. Peat workers played an important, but so far underappreciated, role in the history of energy and industrial growth in the Russian Empire and the Soviet Union. While their bodies served as mediators between the physical environment and the heat, motive power, and electric light consumed by the expanding industrial metabolism, the wages and the goods they received for their hard labor helped sustain their families and improve the modest living conditions in the villages.

The workscapes of the peat industry reveal the power asymmetries and priorities of an industrializing economy, which expanded at the expense of nonhuman life in peatland ecosystems and the health of the workers exploiting these. The women and men who excavated, carried, and stacked peat were much more exposed to the hardships and risks of producing fuel than managers or planners who encountered peat only in the form of abstract numbers. Although the prominent role of female manual workers sets this history apart from cases such as Ireland, where industrial peat extraction was a predominantly male occupation while women were involved in preparing peat for domestic uses, the importance of women is not without parallels in the global history of extractive labor.[94] Despite a shift toward the masculinization of mine work in modern times, female labor has played a crucial role in extractive industries around the world. Yet, regardless of whether it was relegated exclusively to the domestic sphere in mining towns or linked to heavy physical work above ground, this labor was considered "auxiliary" and rendered invisible by contemporaries and historians alike.[95] This was the same in imperial and Soviet Russia, where some of the burdens created by the expanding industrial metabolism fell literally on the shoulders of women. Their hard work in the peat industry was hidden from the users of electricity or manufactured goods and has remained a blind spot in economic and social histories. Amidst an increasing scholarly interest in the material culture of Russia's empires, the case of peat reminds us that material culture in nineteenth and twentieth-century Russia was implicated in extractive regimes which appropriated physical environments and the work of people who, as a rule, had little say in defining the purposes their labor was meant to serve.[96]

Peat workers' lives and work trajectories bring to the fore the overlapping temporalities, modes of production, and agencies (human and nonhuman) shaping Russia's rising fossil economy. Seasonal or temporal labor, which had marked Eurasian growth since the early phases of

[94] On the Irish case, see Proinnsias Breathnach, "Engineering and Re-engineering Earth: Industrialized Harvesting of Ireland's Peatlands and Its Aftermath," in Stanley D. Brunn, ed., *Engineering Earth: The Impacts of Megaengineering Projects* (Dordrecht: Springer, 2011), 429–446, 434 and Toomey, "'A National Resource of Primary Importance'," chapter 2.

[95] Rossana Barragán Romano and Leda Papastefanaki, "Women and Gender in the Mines: Challenging Masculinity through History: An Introduction," *International Review of Social History* 2, no. 65 (2010): 191–230.

[96] Alison K. Smith makes a compelling case for considering the links between St Petersburg's built environment and the limestone from local quarries. See Alison K. Smith, "Lime and Limestone in Eighteenth-Century Russia," in Matthew P. Romaniello, Alison K. Smith, and Tricia Starks, eds., *The Life Cycle of Russian Things: From Fish Guts to Fabergé, 1600–Present* (London: Bloomsbury Academic, 2022), 33–49.

industrial development, endured well into the second half of the twentieth century. And peat workers were no exception. Forest workers were highly mobile throughout the Soviet period. The same is true for the shift workers from Ukraine who, from the 1970s onward, traveled thousands of kilometers between their homes and the oil fields of Western Siberia.[97] The forgotten history of peat not only calls us to recognize cyclical mobilities at the social and geographical margins as constitutive elements of Russia's fossil economy. It also suggests that we think more explicitly about the material premises of industrial growth and its embeddedness in constantly emerging, often unruly more-than-human relationships. Economic histories of Russia consider fossil fuels mostly in terms of their cost or relevance as export commodities without discussing their material role in industries or the socio-ecological impact of their use, while factors such as the weather, the seasons, or the unpredictable agency of unruly matter are either ignored or considered external to the history of industrial expansion.[98] Although peat extraction sites may seem negligible within larger narratives of Eurasian growth, the forgotten margins of Russia's fossil economy were focal points of the intertwined exploitation of people and nature that underpinned its expansive trajectory from the late nineteenth century. Bringing our histories "down to earth," we gain a more comprehensive but, at the same time, much messier picture of Russia's fossil economy, one marked by a myriad of social and material entanglements between superficially unrelated places, people, and things.

[97] Burkush, "On the Forest Front"; Alexander Etkind, Yevhenii Poliakov, and Bohdan Shumylovych, "Ukrainian Labor and Siberian Oil in the Late Soviet Empire," *Journal of Soviet and post-Soviet Politics and Society* 6, no. 2 (2020): 241–280. Construction workers were highly mobile in the late Soviet period, too, as shown in Siegelbaum and Moch, *Broad Is My Native Land*, 87–95.

[98] Most quantitative studies do not consider fossil fuels (or energy more broadly) central to the history of Eurasian industrial growth, though some mention resource depletion and wastefulness in the context of declining growth rates in the late Soviet period. See Robert C. Allen, "The Rise and Decline of the Soviet Economy," *Canadian Journal of Economics* 34, no. 4 (2001): 859–881, 876–878; Philip Hanson, *The Rise and Fall of the Soviet Economy: An Economic History of the USSR from 1945* (Oxon: Routledge, 2003), 132–140. A recent overview of the relationship between state and market in the history of Russian industrialization mentions energy only in relation to production cost and export revenue. Andrei Markevich and Steven Nafziger, "State and Market in Russian Industrialization, 1870–2010," in Kevin Hjortshøj O'Rourke and Jeffrey Gale Williamson, eds., *The Spread of Modern Industry to the Periphery since 1871* (Oxford: Oxford University Press, 2017), 33–62.

4

Transformation

In November 1957, 40 years after the Bolshevik takeover, things looked bright in the region of Shatura: "Where now there are comfortable settlements and expansive peat-extraction sites, before the Great October, there lay marshy impenetrable swamps and centuries-old forests." For I. Alekseev, who published these lines in *Za Torf*, the weekly newspaper of the Shatura Peat Trust, the transformation of nature in the Meshchera Lowland proved the progress which Lenin and the communist party had brought to the inaccessible and underdeveloped environments to the east of the city of Moscow. Not only had Soviet power made it possible to tap into the natural riches of central Russia's extensive wetlands, it also provided peat workers with places to live in decency and with some basic comforts. Within the structures of Shaturtorf alone, some 23 schools, 21 kindergartens, 13 nurseries, 9 hospitals, 22 ambulances, and just under 200,000 square meters of living space powerfully testified to the legacy of the revolution and its vision of social progress.[1]

Alekseev's article was just as true as it was false. Readers at the time would have easily recognized the narratives it invoked. Lying at the core of the Soviet tale of origin, the "Great October" had been evoked since the 1920s in annual celebrations, history books, street names, political speeches, and cultural production to underscore the legitimacy of Soviet power. With Khrushchev's 1956 "Secret Speech," reference to Leninist principles became a crucial element in the rhetoric and politics of

[1] I. Alekseev, "Tam, gde byli neprokhodimye bolota," *Za Torf* (November 14, 1957): 1.

De-Stalinization.[2] Alekseev clearly overestimated the role of Lenin and the October Revolution in the appropriation of the peatlands east of Moscow. As previous chapters have discussed, commercial peat extraction in the Meshchera originated in the imperial period. Even the idea to convert peat into electricity, which gained such prominence in the GOĖLRO-plan, predated 1917. Still, Alekseev's account was more than a formulaic exercise in Soviet myth-making during the optimistic years of the Thaw. In 1957, the Shatura region had changed dramatically since the time of the Revolution. After the Bolshevik leadership had decided to rely on the peatlands of central Russia in their electrification efforts, extractive activities reshaped the landscape in profound and lasting ways. Peat trusts, territorial organizations uniting several peat-extracting companies, removed millions of tons of peat from Meshchera's wetlands every year, turning complex and diverse ecosystems into industrial landscapes which served the needs of power utilities and users of combustible fuel. At the same time, the peat industry provided basic civilian infrastructure for workers who flocked to the extraction sites in ever-growing numbers from the early 1920s. The result was ambivalent. Gradually degraded from an ecological point of view, the peatlands around Shatura became increasingly habitable for the people who exploited them.

Geographers would call the transformation of wetland environments through resource extraction, infrastructure, settlement, and the endowment of the area with symbolic significance "place-making," a process by which abstract, indistinct *space* turns into a value- and meaning-bearing *place*. Places are not static but continuously made and remade, sometimes even unmade, through activities that alter the relationships between the social and the material worlds and shape people's emotions toward and understandings of their surroundings.[3] For historians of the Russian Empire and the Soviet Union, space and place are familiar categories. Russian rule in Eurasia was closely tied to the exploration, integration, mastery, and transformation of space through imperial technologies and epistemologies, settler colonialism, bureaucratic rule, and cultural

2 Frederick C. Corney, "Rethinking a Great Event: The October Revolution as a Memory Project," *Social Science History* 22, no. 4 (1998): 389–414; Polly Jones, *Myth, Memory, Trauma: Rethinking the Stalinist Past in the Soviet Union, 1953–70* (New Haven: Yale University Press, 2013), 21, 32–33.

3 Tim Cresswell, *Place: A Short Introduction* (Malden, MA: Blackwell Publishing, 2004), 5–10; Robert David Sack, "The Geographic Problematic: Empirical issues," *Norsk Geografisk Tidsskrift – Norwegian Journal of Geography* 55, no. 3 (2001): 107–116.

erasure.[4] Developmentalist phantasies, violence, dispossession, and displacement have been deeply woven into Eurasian histories of place. During the collectivization campaign of the early 1930s and the Second World War, millions of women and men were coerced to leave their homes and sent to unknown environments, where the Soviet government sought to extract resources or develop industries. As GULAG inmates or "special settlers," they endured years of exploitation, usually without even the most basic provisions. Of those who survived, many tried to return to their places of origin once this became possible. Others slowly, often experiencing discrimination and prejudice, began to make a home in their places of exile as life following Stalin's death became more stable and Soviet identity discourses emphasized welfare, consumption, and the war victory while silencing the population's traumatic past. Whether in the mining towns of the Russian far north, sometimes very close to former prison camps, or in the Kazakh steppe, where Stalin had deported members of non-Russian ethnic minorities and where thousands resettled during Khrushchev's "Virgin Land Campaign," people over time articulated local identities. Personal involvement in shaping these places, be it through work in extraction and industry or activities such as gardening or urban greening, often underwrote the emergence of place-based feelings of attachment and connection in the Soviet state.[5]

[4] Karl Schlögel, ed., *Mastering Russian Spaces: Raum und Raumbewältigung als Probleme der russischen Geschichte* (Munich: Oldenbourg, 2011); Mark Bassin, Christopher Ely, and Melissa K. Stockdale, eds., *Space, Place, and Power in Modern Russia: Essays in the New Spatial History* (DeKalb: Northern Illinois University Press, 2010). For surveys of the scholarship on the "spatial turn" in Russian and Soviet History, see Nick Baron, "New Spatial Histories of Twentieth Century Russia and the Soviet Union: Surveying the Landscape," *Jahrbücher für Geschichte Osteuropas* 55, no. 3 (2007): 374–400; Frithjof Benjamin Schenk, "Das Paradigma des Raumes in der Osteuropäischen Geschichte," *zeitenblicke: Online Journal für die Geschichtswissenschaften* 6, no. 2 (2007), www.zeitenblicke.de/2007/2/schenk; Malte Rolf, "Importing the 'Spatial Turn' to Russia: Recent Studies on the Spatialization of Russian History," *Kritika: Explorations in Russia and Eurasian History* 11, no. 2 (2010): 359–380.

[5] Alan Barenberg, *Gulag Town, Company Town: Forced Labor and Its Legacy in Vorkuta* (New Haven: Yale University Press, 2014), 222–230; Alla Bolotova, "Loving and Conquering Nature: Shifting Perceptions of the Environment in the Industrialised Russian North," *Europe-Asia Studies* 64, no. 4 (2012): 645–671; Alla Bolotova and Florian Stammler, "How the North became Home: Attachment to Place among Industrial Migrants in the Murmansk Region of Russia," in Lee Huskey and Chris Southcott, eds., *Migration in the Circumpolar North: Issues and Contexts* (Edmonton: CCI Press, 2010), 193–220; Michaela Pohl, "The 'Planet of One Hundred Languages': Ethnic Relations and Soviet Identity in the Virgin Lands," in Nicholas B. Breyfogle, Abby Schrader, and Willard Sunderland, eds., *Peopling the Russian Periphery: Borderland Colonization in Eurasian History* (London: Routledge, 2007), 238–261.

Usually associated with the geographical fringes of Russia's empires, resource extraction, migration, and place were closely entangled in regions near the political and industrial centers, too. Different groups of actors contributed to remaking space into place in the Meshchera Lowland. Planners and engineers designed and built the infrastructure which facilitated the peat industry at the local level and integrated peatlands into transregional flows of energy, things, and people. Workers formed new communities. Journalists and historians ascribed cultural meaning to Meshchera's peatlands. It took decades for these processes to translate into stable ideas of place. In fact, up until the second half of the twentieth century, many peat extraction sites and their surroundings remained transient spaces, as most workers stayed there only for the peat season. The region acquired more permanent features once the shift to milled peat extraction from the 1950s drastically reduced the demand for seasonal labor, while increasing state investment in civilian infrastructure and social welfare following Stalin's death in 1953 notably improved workers' living conditions.

The transformation of central Russia's physical and imagined geographies during a century of large-scale peat extraction was driven by industrial interest, perpetuated social difference, and reproduced hegemonic narratives. But this process was also shaped from below, as peat workers began to appropriate their continuously evolving environments in their own, often messy ways. Peat extraction sites and their surroundings were microcosms of the conflicting trends marking the rise of Russia's industrial metabolism which brought ecological disruption and engendered new forms of social cohesion. Exploring the intersection between resource-extraction and the making of place, this chapter foregrounds the multiple meanings of peat in Russia's fossil economy. Peat was more than a fuel at the service of industries and power stations. In fact, it also generated new place-based feelings of belonging, which drew from larger tales of progress and patriotism as much as from a sense of attachment within communities which emerged through the intertwined material and cultural histories of extraction and construction, production and reproduction, personal experience and the writing of local history.

STRAIGHT LINES ACROSS IRREGULAR TERRITORY

Before the schools, kindergartens, and apartment blocks that Alekseev listed with such accuracy came to signify the advent of late Soviet normality, Meshchera's peatlands underwent massive transformations that incorporated them into the expanding fossil economy. Wherever peat was extracted at an industrial scale, infrastructures of various kinds, sizes, and materialities penetrated the land. Mediators between humans and peatlands, canals, cables, and rails profoundly reordered wetland environments in central Russia, subjugating them to the imperative of extraction. The anthropologist Tim Ingold has argued that "the straight line has emerged as a virtual icon of modernity, an index of the triumph of rational, purposeful design over the vicissitudes of the natural world."[6] The changing geography of central Russian peatlands underscores this point. The infrastructures to extract and transport peat inscribed visions of order, regularity, and predictability into environments which, at least from the point of view of outsiders, appeared to be chaotic *space* which had to be mastered before becoming a meaningful and useful *place*. These infrastructures, moreover, were not socially neutral. Promoting movement and spatial cohesion on the one hand, canals, railways, and power lines also fragmented previously connected territory, created new or reinforced existing social hierarchies, and established boundaries that were difficult to overcome once in place.[7]

Drainage canals were usually the first physical markers indicating the arrival of modern notions of productive land use in peatland environments. The presence of canals in the landscapes of central Russia predated the large-scale extraction of peat as an industrial fuel. Drawing boundaries between water and land, canals were a tool of spatial control, making it easier for people to move across and systematically utilize wetlands. Noble landowners had recognized this potential and locally practiced land drainage since the late eighteenth century, mainly to improve conditions for farming and forestry. More systematic interventions in wetland ecosystems began in the second half of the nineteenth century. The German gentry in the Baltic provinces promoted drainage and experimented with mechanized peat extraction since the 1850s. The central administration in Saint Petersburg became invested in large-scale wetland reclamation during the early 1870s when the state-led Western Expedition for the Drainage of the Swamps in Polesia began to drain the land along the Pripyat River in the empire's Western borderlands and soon after that in the central Russian provinces of Vladimir, Moscow, Tver, and Ryazan. In the decades before the First World War, the

[6] Tim Ingold, *Lines: A Brief History* (London: Routledge, 2007), 152.

[7] On writing the history of infrastructure, see Dirk van Laak, "Infra-Strukturgeschichte," *Geschichte und Gesellschaft* 27, no. 3 (2001): 367–393.

Department of Land Improvement carried out substantial reclamation works across the empire. The maps, photographs, and written reports produced by the imperial authorities normalized the idea that wetlands should be tamed through ambitious engineering.[8]

Late imperial reclamation works had a lasting legacy. In fact, in some parts of central Russia, they laid the groundwork for the canal infrastructure, which would support large-scale peat extraction during the twentieth century. Land surveys and canal construction initially focused on state-owned estates (*kazennye dachi*) to turn these into sources of income. In the Meshchera, Tsarist officials turned their eyes to the marshy lands of the Radovitskaia, Shaturskaia, Tugoletskaia, and Petrovsko-Kobelevskaia estates.[9] During the Soviet period, many of these lands were allocated to the companies of the Shatura Peat Trust, some of which would carry the memory of the former imperial estates in their names. Although most drainage interventions before 1917 aimed at exploiting the surface of peatlands, extractivist motives were not absent. As Chapter 1 has demonstrated, rising fuel prices made the lease of state-owned wetlands to industrial companies for fuel extraction a welcome source of fiscal revenue in the final decades of Tsarist rule. Shortly before the First World War, the Department of Land Improvement considered draining a state-owned peatland in the Vladimir province to divide it into parcels that industrial businesses would lease to access cheap fuel.[10] In the case of the state-owned peatland Tugolesskii Bor at the north-western edge of the Ryazan province, state actors equally endorsed peat extraction as a profitable mode of land use. By 1913, some 100 versts of canals, dug with the help of peasants from surrounding villages, permeated the land (Figure 4.1).[11] The First World War put plans to convert the wetland into an extraction site on hold, but the region did not lose its attractiveness as a peat deposit. Shaturtorf's company Tugolesskii Bor exploited the area from the 1930s until the late Soviet period.[12]

[8] Ruuskanen, "The Emergence of Baltic *Moorkultur*"; Bruisch, "The State in the Swamps."

[9] *Kratkii ocherk rabot ėkspeditsii po osusheniiu bolot v Tsentral'nom raione* (Saint Petersburg: Zapadnaia Ėkspeditsiia po osusheniiu bolot, 1896), 4–5.

[10] "Gidrotekhnicheskie raboty v tseliakh rasshireniia i uluchsheniia proizvoditel'noi ploshchadi i pri zemleustroistve," in GUZiZ, *Ezhegodnik Otdela zemel'nykh uluchshenii, 1912* (Saint Petersburg: Tipografiia V. F. Kirshbauma, 1913), 36–38.

[11] "Gidrotekhnicheskie raboty v Evropeiskoi Rossii v tseliakh rasshireniia i ulushcheniia proizvoditel'noi ploshchadi," in GUZiZ, *Ezhegodnik Otdela zemel'nykh uluchshenii, 1913* (Petrograd: Tipografiia V. F. Kirshbauma, 1914), 95–96, 126.

[12] "Godovoi otchet stroitel'stva 'Tugolesskii bor' za 1936g.," RGAĖ f. 7969, op. 1, d. 162, ll. 3–20.

FIGURE 4.1 Canal construction in the Meshchera. Source: GUZiZ, *Ezhegodnik Otdela zemel'nykh uluchshenii. 1913. Ch. 1* (Petrograd: Tipografiia V. A. Kirshbauma, 1914), 72.

Railways similarly tied peatlands to the flows of matter and energy within the expanding fossil economy. An important tool of spatial consolidation in the Russian Empire, railways linked metropoles to distant regions such as Central Asia or Siberia, but they also integrated the social, political and economic space in European Russia.[13] In the twentieth century, the Soviet peat industry developed an expansive network of rail tracks to move peat between extraction and storage sites and to connect extraction fields, power stations, and industrial peat users. Most of them were for narrow-gauge railways, as these could be built much faster than regular track-width railways. Their smaller size and lower weight also made them much better suited for the dense vegetation and the soft ground of the marshy land. In the 1980s, narrow-gauge railway tracks spanning a total length of 500 kilometers crisscrossed the territory managed by the Shatura Peat Trust.[14] Peat's significance as a fuel for the power industry left a mark on the geographies of interregional transport, too. During the 1930s, Mosėnergo, the organization in charge of Moscow's power supply, was closely involved in constructing a rail line between the towns of Shatura and Orekhovo-Zuevo, while the Shatura and Orekhovo-Zuevo Peat Trusts contributed labor and other resources to the building works. For Mosėnergo, the railway connection was a means to secure steady fuel supplies for the Shatura and Ėlektroperedacha power utilities, whose transmission lines beamed electric power to the Soviet capital.[15] Narrow-gauge railway tracks and freight cars filled with peat never became part of the iconography of Soviet electrification, which has centered around complex and large technologies. An important part of the intricate chain linking peatlands, power plants, and electricity consumers, however, these simple technologies were essential for the generation of electric power.

Whereas canals and narrow-gauge railways penetrated wetlands' interiors, power plants and transmission cables connected them to places further apart, creating what the historian Christopher F. Jones called "landscapes of intensification," brought about and continuously reproduced by "a positive feedback loop between supply and demand."[16] Once regional utilities started operating, central Russia's peatlands, which local entrepreneurs had exploited since the late nineteenth century, became part

[13] Frithjof Benjamin Schenk, *Russlands Fahrt in die Moderne: Mobilität und sozialer Raum im Eisenbahnzeitalter* (Stuttgart: Franz Steiner Verlag, 2014), 37–118. For an environmental perspective on late imperial railways, see Keating, *On Arid Ground*, 35–44.

[14] Shatura Peat Trust, Annual Report (1935), RGAĖ f. 7969, op. 1, d. 123, l. 207; Shatura Peat Trust, Annual Report (1986), GARF f. A-632, op. 9, d. 475, ll. 69–70.

[15] "Otchet stroitel'stva zhelezno-dorozhnoi vetvi Orekhovo-Shatura (1937)," RGAĖ f. 7969, op. 1, d. 202, ll. 1–13.

[16] Christopher F. Jones, *Routes of Power: Energy and Modern America* (Cambridge, MA: Harvard University Press, 2014), 9.

of a spatially integrated energy landscape. Mediators between fuel extraction and energy consumption, the power plants Ėlektroperedacha, built near the town of Bogorodsk shortly before the First World War, and Shatura, the poster child of the GOĖLRO, drastically expanded the geographical reach of peat. Converting Meshchera's peat into electricity, which then traveled on to the city of Moscow and industries in the area, Ėlektroperedacha and Shatura embodied a new energy regime, one in which peatlands functioned as long-time providers of fuel serving the demands of distant places which had no immediate link to the material origins of electrical power.

The consolidation of space through infrastructure generated friction on the ground. Designed to facilitate the movement of matter or energy, infrastructure regularly interfered with mobility patterns and land uses in rural areas. Before the 1917 Revolution, system builders who claimed to act in the name of the "public good" regularly clashed with private landowners or peasant communities over the necessity, the ownership, or the location of railway lines and over the compensation that landowners could expect if their land was given to the railways.[17] The appropriation of peatlands had similar effects. For instance, digging drainage canals was a far more ambivalent action than the notion of improvement suggested. Drainage canals could provide economic opportunities for land-owning elites and peasants, but they became a source of frustration when built through land that locals saw and used as integrated space. Take the 1898 case of a forester from the province of Vladimir, who appealed to the Department of Land Improvement in Saint Petersburg requesting that five bridges be constructed. The forester had good reasons not to regard this as a local matter. The need for these bridges had arisen because drainage canals dug upon the Department's decision had cut through paths that the local population needed for cattle driving and the transportation of wood. His troubles were no exception. In other places, the central administration had built bridges over newly dug drainage canals but did not maintain them properly, making them unusable over time.[18] Cables could equally cause local discontent. In 1912, the managers of Russia's first peat-fueled power plant Ėlektroperedacha faced a long conflict about the course of the power line meant to bring electricity to the city of Moscow. Concerned about their control over the land,

[17] Pravilova, *A Public Empire*, 97–98.

[18] "O sostoianii mostov, vodospuskov v Moskovskoi, Tverskoi, Riazanskoi gub. 1900 g.," RGIA, f. 424, op. 1, d. 271, ll. 6–7.

members of the zemstvo refused to grant the engineers permission to build the line along what would have been the most direct connection, the Nizhny Novgorod road. The engineers subsequently convinced local peasant communities to accept power poles being placed on their land. Robert Ė. Klasson, one of Ėlektroperedacha's chief engineers, who was closely involved in these negotiations, later commented that the episode taught him that private property was a major obstacle to any efforts to electrify a country on a large scale.[19]

The abolition of private ownership of land and resources following the 1917 Revolution made implementing infrastructural projects easier, but this did not imply that new infrastructure served everybody equally. From its inception, Soviet electrification prioritized industrial and urban needs. The GOĖLRO-plan projected three high-voltaic transmission lines radiating from Shatura in different directions: one would connect Shatura directly with Moscow, another would run to the Soviet capital via the manufacturing centers Egor'evsk and Kolomna, and a third would lead to the town Pokrov from where electricity would move in a north-eastern direction toward the city of Vladimir.[20] The planners projected that only villages within a radius of five versts from the Shatura Power Station could be expected to receive electric energy in the near future.[21] Despite the propagandistic noise around the light bulb illuminating peasant huts, the rural–urban divide remained embedded in the Soviet energy system for quite some time.[22] A map in the 1933 Atlas of the Moscow region documenting the geographical reach of power utilities indicates that the areas surrounding the Shatura and Ėlektroperedacha power plants had not yet been electrified. While Meshchera's peat brought to life lamps and machines in faraway places, its rural communities had to wait until they could finally access the electricity generated from the peat riches in their vicinity (Figure 4.2).

I got a glimpse of the unequal spatial distribution of energy access in central Russia when visiting the Shatura region in the summer of 2019. In a largely abandoned workers' settlement that had once housed Shaturtorf's workers, I met two men who ran an informal repair service for cars and lorries surrounded by crumbling wooden barracks and dense

19 "Klasson's memoirs (early 1920s)," RGAĖ f. 9508, op. 1, d. 9, l. 17.
20 Simonov, *Razvitie ėlektroėnergetiki*, 270.
21 "Doklad o razvitii tekstil'noi promyshlennosti v Tsentral'noi Rossii (1920)," GARF f. 5208, op. 1, d. 10, l. 173.
22 Coopersmith, *Electrification of Russia*, 236–244.

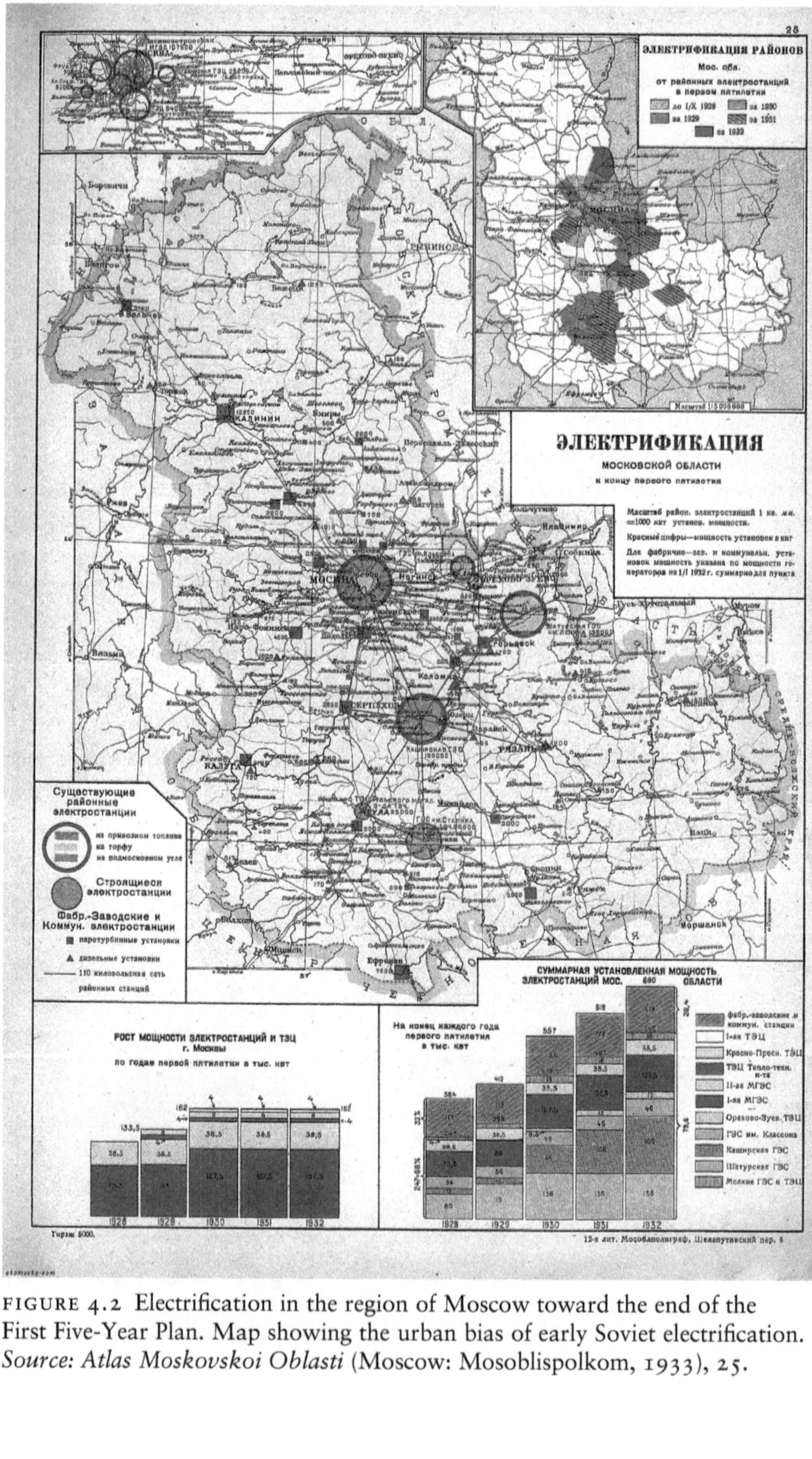

FIGURE 4.2 Electrification in the region of Moscow toward the end of the First Five-Year Plan. Map showing the urban bias of early Soviet electrification. *Source: Atlas Moskovskoi Oblasti* (Moscow: Mosoblispolkom, 1933), 25.

vegetation reconquering the space after most people had moved away. We were chatting less than five kilometers as the crow flies from the Shatura Power Plant, the power line running directly above our heads. When I asked the men about their power source, they responded "We steal electricity" (*ėlektroėnergiiu voruem*) and pointed to the hardware that helped them draw electricity from the transmission line.[23] It is not unlikely that the settlement had been connected to the power network before peat extraction ceased and most workers left. Still, the place exemplified how inequalities embedded in the spatial design of the grid force people on the ground to find their own ways to make ends meet. Technologies of straight lines, as the historian Kate Brown once put it, "helped to fix social relations in space."[24]

Such inequalities also shaped the sphere of transport. Getting around was much more difficult for humans than it was for peat. When workers wanted to go places beyond their settlement or workplace, they could find themselves excluded from using transport infrastructure built in the vicinity of their homes. For example, consider the short railway section between the Sazonovo and Pilevo settlements at the administrative border of the Moscow and Ryazan oblasts. Realized with resources from the Ministry of Fuel Industry, the 32-kilometer track was built to transport peat from the extensive extraction fields near Pilevo to the Shatura Power Plant. After its completion in 1976, the line was handed over to the Ministry of Transport, which delegated responsibilities for the line to the Moscow Railways, an organization technically independent from the interests of the peat industry.[25] Still, other than initially intended, regular passenger trains never ran on this route. Until its suspension in the 2000s, the line served freight trains almost exclusively.[26] For the people living in Radovitskii, a workers' settlement housing Shaturtorf's workers, the subordinate priority of their needs must have been a cause of ongoing

[23] Conversation with the author, Shaturtorf, June 9, 2019.

[24] Kate Brown, "Gridded Lives: Why Kazakhstan and Montana are Nearly the Same Place," *The American Historical Review* 106, no. 1 (2001): 17–48, 46. For an excellent discussion of electricity theft, see Diana J. Montaño, "Ladrones de Luz: Policing Electricity in Mexico City, 1901–1918," *Hispanic American Historical Review* 101, no. 1 (2011): 35–72.

[25] "O prieme v ėkspluatatsiiu zheleznodorozhnoi linii torfopredpriiatiia Meshcherskoe Shaturskogo proizvodstvennogo ob-edineniia Ministerstva toplivnoi promyshlennosti RSFSR (1975–76)," GARF f. 259, op. 46, d. 6931, ll. 1–3, 35.

[26] I am relying here on the information by local rail enthusiasts who have collected plenty of historical and visual material about the line. "Sazonov-Pilevo" (https://obolev.livejournal.com/13791.html); "Zheleznodorozhnaia liniia Sazonovo-Pilovo" (https://kirillfedorov4.livejournal.com/13483.html). Accessed June 7, 2024.

frustration. Even though cargo trains loaded with peat regularly passed by on their way to Shatura, they had to take two buses to reach the same destination. The journey to the town, where workers dealt with bureaucratic issues or purchased goods unavailable in local stores, could take two and a half to three hours. A passenger train on the line from Sazonovo to Pilevo with a stop in Radovitskii would have made this journey significantly faster.[27]

Manifestations of social hierarchies, technologies of the straight line were an ecological imposition as well, reducing complex and diverse wetland ecosystems to a two-dimensional canvas that experts and planners filled with their vision of controlling nature. Reclamation manuals promoted standardized drainage systems, which organized wetlands in a pattern of rectangular extraction fields bordered and divided by canals and ditches.[28] These ideas directly translated into the spatial setup of peat companies, particularly once milled peat proliferated in the decades following the Second World War. In 1969, after years of planning and preparation, Shaturtorf commissioned its new branch Petrovskoe to extract milled peat from three deposits just a few kilometers east of Shatura. Closely mirroring contemporary textbook depictions of such sites, Petrovskoe's drainage system consisted of parallel magisterial canals with a 500-meter distance between them, which channeled the water into surrounding rivers and lakes. Smaller ditches running perpendicularly to the main drains crossed the land every 20 or 40 meters.[29] These changes had a lasting impact. Drainage and peat extraction were acts of terraforming geometry whose imprints on the Earth's surface were so profound that satellites can still detect these regular structures from outer space (Figure 4.3).

Once the infrastructure was in place, peat trusts found themselves in a constant struggle to maintain the logic of the straight line, as peatlands resisted human attempts to make them fully legible and predictable. Rails sank into the swampy ground; water leached through canal walls or bypassed sluices; severe frost forced peat trusts to interrupt railway transport.[30] Trusts' difficulties in protecting these systems from natural

[27] "Zheleznodorozhnaia liniia Sazonovo – Pilevo" (https://railway-archive.studio-petukh.ru/article?name=06/pilevo.html). Accessed June 7, 2024.

[28] V. V. Shablinskii, *Osushenie torfianykh mestorozhdenii* (Moscow: Gosudarstvennoe ėnergeticheskoe izdatel'stvo, 1953), 50–54.

[29] "Akt priema v ėkspluatatsiiu torfopredpriiatiia 'Petrovskoe' Shaturskogo torfotresta 17 dekabria 1969 goda," GARF f. A-632, op. 2, d. 287, ll. 9–11.

[30] See, for instance "Otchet stroitel'stva zhelezno-dorozhnoi vetvi Orekhovo-Shatura (1937)," RGAĖ f. 7969, op. 1, d. 202, l. 7; "Otchet stroitel'stva zhelezno-dorozhnoi

FIGURE 4.3 Abandoned peat extraction sites formerly exploited by the Petrovskoe Peat Company northeast of Shatura. Created with Google Earth / 2024 Maxar Technologies.

decay, aggravated by poor maintenance or the lack of spare parts for rails, trains, and pumps, increased over time. From the 1970s, the underutilization (*neosvoenie*) of the railways, as the problem was framed in bureaucratic terms, became an enduring concern for the authorities. Shaturtorf, for instance, repeatedly failed to deliver the assigned amount of peat to the Shatura Power Plant because its wagons, rails, or other transport-related machines were dysfunctional.[31] That such issues filled

vetki Orekhovo-Shatura (1938)," RGAĖ f. 7969, op. 1, d. 239, ll. 1–3; "Protokoly, otchety, akty i dr. materialy obsledovaniia setei osusheniia, proizvodstvennogo i protivopozharnogo vodosnabzheniia torfopredpriiatii (1953)," RGAĖ f. 7969, op. 2, d. 672, ll. 69–100; "Reshenie Glavnogo Upravleniia torfianoi promyshlennosti po godovomu otchetu Shaturskogo gosudarstvennogo torfianogo tresta za 1965 god," GARF f. A-632, op. 2, d. 39, l. 74; "O podgotovke zheleznodorozhnogo transporta uzkoi kolei proizvodstvennykh ob-edinenii po dobyche torfa k osenne-zimnim perevozkam 1980-1981gg.," GARF, f. A-632, op. 9, d. 252, 59–63.

31 See, for instance "O postavke torfa na Shaturskuiu GRĖS No. 5 imeni V. I. Lenina (1974)," GARF f. A-632, op. 2, d. 537, ll. 7–8; "Reshenie Ministerstva toplivnoi promyshlennosti RSFRS po godovomu otchetu Shaturskogo proizvdostvennogo ob-edineniia po dobyche torfa (1975)," GARF A-632, op. 6, d. 52, ll. 88–89. See also "O neosvoenii proizvodstvennymi ob-edineniiami vagonov v mae s. g." (1983), GARF f. A-632, op. 9, d. 374, ll. 35–38.

page after page in the archival record illustrates a fundamental problem haunting the peat industry and the Soviet economy at large. As productivist ambitions trumped the need for investments in service and repair, nature's ability to undermine or disrupt human plans was constantly reinforced.

LANDSCAPES OF REPRODUCTION

Infrastructure was only one marker of how the use of peat for heat and power transformed central Russia's wetland environments. Since peat fuel production remained a labor-intensive process well into the second half of the twentieth century, Meshchera's peatlands began to sustain more people than at any other time in history. From the late nineteenth century, workers' settlements (*rabochie poselki*) started dotting the land wherever peat works were set up. These settlements were a type of company town, a place where a company acts as both employer and provider of goods and services.[32] Designed to organize productive and reproductive activities in spatial proximity, company towns played an essential role in industrializing countries worldwide, including the Russian Empire and the Soviet Union.[33]

Marrying traits of urban and rural ways of life, workers' settlements attached to the peat industry were much smaller than the sometimes massive industrial agglomerations known as "monotowns." Production and reproduction constantly overlapped in peat workers' settlements, but their relationship changed profoundly over time. As long as the workforce remained predominantly seasonal, most peat workers stayed in rudimentary accommodation designed for temporary use. From the mid-1950s, when seasonal labor became less prominent, workers got access to more permanent housing, leisure facilities, and welfare services while also finding ways to make the settlements their own. It was in the social and

[32] Marcelo J. Borges and Susana B. Torres, "Company Towns: Concepts, Historiography, and Approaches," in Marcelo J. Borges and Susana B. Torres, eds., *Company Towns: Labor, Space, and Power Relations across Time and Continents* (New York: Palgrave Macmillan, 2012), 1–40.

[33] Volodymyr Kulikov, "Industrialisation and Transformation of the Landscape in the Donbass from the late Nineteenth to the early Twentieth Century," in Martin Zückert and Heidi Hein-Kirchner, eds., *Migration and Landscape Transformation: Changes in Central and Eastern Europe in the 19th and 20th Century* (Göttingen: Vandenhoeck & Ruprecht, 2016), 57–81; Mark Meerovich, *Gradostroitel'naia politika v SSSR 1917–1929: Ot goroda-sada k vedomstvennomu rabochemu poselku*, 2nd ed. (Moscow: NLO, 2018).

physical environments of these settlements that peat workers would come to experience the Soviet enterprise as "a moral as well as a productive community," as the sociologist Sarah Ashwin once put it.[34]

Some simple welfare provisions for peat workers existed as early as the late imperial period, when industrial companies began extracting peat to fuel their factories. The Nikolskaia cotton mill, for instance, set up a small medical station in a heated barrack close to a peat extraction site because the mill's main hospital was too difficult to reach from there.[35] Measures like this reflected a broader trend toward the expansion of welfare work in late imperial businesses, which invested in social infrastructure to secure their labor supply.[36] Because their role was considered tangential to the core operations of large industrial mills, peat workers did not rank highly in such schemes, however. When state institutions, in response to a massive strike that swept through central Russian textile mills during the 1880s, monitored workers' labor and living conditions, peat workers' marginal position within factory welfare became obvious. A set of formal instructions issued by the governor of the Moscow province indicates that decent housing and sanitation were nothing that peat workers could count on. Prohibiting that peat workers live in dugouts and cellars, the instructions prescribed that dormitories have sturdy walls and roofs, ovens, bunks or beds, and some dedicated space for laundry. Companies were also expected to ensure basic medical services within a distance of ten verst from extraction sites.[37] A few decades later, the Soviet state followed a similar approach to introduce housing and sanitation norms. In the early 1920s, Soviet authorities even defined the amount of air (9.66 cubic meters) and space (4 square meters) a peat worker in shared accommodation was entitled to.[38] Given that the authorities saw the need to spell out these terms, workers' everyday lives probably looked much bleaker.

Social inequality was deeply embedded in the design of workers' settlements. Administrators, engineers, and skilled workers were

34 Sarah Ashwin, "Endless Patience: Explaining Soviet and Post-Soviet Social Stability," *Communist and Post-Communist Studies* 31, no. 2 (1998): 187–198, 193.

35 Potkina, *Na olimpe delovogo uspekha*, 163.

36 Volodymyr Kulikov, "Necessity or Luxury? Welfare Work in the Company Towns of the Russian Empire," *Jahrbuch für Wirtschaftsgeschichte* 60, no. 2 (2019): 449–472.

37 "Obiazatel'nye sanitarnye postanovleniia po ustroistvu i soderzhaniiu zhilykh i drugikh pomeshenii dlia rabochikh na torfianykh bolotakh (1892, 1989)," RGAĖ f. 478, op. 7, d. 2671, ll. 56–57.

38 "Tsirkuliar Otdela Okhrany Truda NKT SSSR o zhilishchnykh usloviiakh rabochikh na torfo-razrabotkakh," *Ėkonomicheskaia Zhizn'* (January 27, 1924), 7.

privileged when it came to housing and other services in company-built settlements. The revolution did not bring any substantial change to this.[39] Like many early Soviet planners and engineers, Alexander V. Vinter, who was one of the chief engineers involved in constructing the Shatura Power Station, dreamed of housing workers in places reminiscent of English garden cities.[40] When Shatura grew into a town following the commissioning of the power station in the 1920s, some elements of the garden city were incorporated into the town's layout.[41] However, the settlements for workers who were not involved in running the prestigious utility but were only temporarily employed to extract peat, remained transient spaces with poor living conditions and services. In the 1930s, Shaturtorf's workers lacked things as simple as chairs, tea kettles, or cups. Their barracks were unsuited to the humid environments of peatlands and were frequently damaged. Not all of them could be heated, and broken windows or doors often went without repair after workers returned to their villages for the winter.[42]

Shaturtorf's administration does not seem to have had the means or the desire to address these issues. In 1950, the Trade Union of Peat Workers lamented that the trust's hostels lacked sinks, stools, and storage, and that workers slept in old, rusty beds that they sometimes had to share. In one of Shaturtorf's enterprises, up to 12 people lived together in a room where they also prepared their food because they had no access to a kitchen.[43] Peat workers lived and worked in conditions shaped by the nature of the peatlands and their work. But their daily deprivations closely mirrored those facing most of the working population during Stalinism. The rapid pace of industrialization since the early 1930s went along with a general neglect of the Soviet population's basic needs. Despite some modest improvements, Soviet workers still ate poorly and lived in overcrowded settings, often without running water, sewage, or central heating in the

[39] Kulikov, "Necessity or Luxury?," 457–465; Meerovich, *Gradostroitel'naia politika*, 252–253.

[40] "Vospominaniia o Shature," GARF f. 7952, op. 3, d. 459, ll. 3–10. On the ideal of the garden city in early Soviet urban planning, see Stites, *Revolutionary Dreams*, 191–196.

[41] The image of the garden city is still used to market Shatura as a destination for domestic tourists. See https://welcome.mosreg.ru/stories/lenin-muzei-i-gorod-sad-progulka-po-sature. Accessed July 31, 2024. For a glimpse of Shatura's early Soviet architecture, see https://dmitry-sasin.livejournal.com/492096.html. Accessed July 31, 2024.

[42] Shatura Peat Trust, Annual Report (1934), RGAĖ f. 7969, op. 1, d. 77, l. 193; Shatura Peat Trust, Annual Report (1937), RGAĖ f. 7969, op. 1, d. 213, l. 37.

[43] "Resolution of the Trade Union of Workers of the Peat Industry (January 1950)," RGAĖ f. 7969, op 2, d. 641, ll. 30–34.

years following the Second World War.[44] For peat trusts like Shaturtorf, keeping daily provisions for workers to a minimum made economic sense. Up until the postwar decades, life in the Soviet countryside was so dire that getting some extra income remained attractive for villagers despite the harsh work and modest life that awaited them at the peat works. As peat trusts did not fear any immediate shortage of working bodies, maintaining investments in social infrastructure at a low level was a means to keep their labor cheap.

Food, another marker of the relationship between production and reproduction in workers' settlements, was of equally low quality and a recurrent cause of workers' discontent. Sarcastic and angry comments about how little and poorly they ate filled entire verses in peat workers' folklore: "In the swamp life is joyful / Drinking tea, but nothing to eat . . ." And the song went on:

In the M . . . swamp	На М . . . болоте
Porridge they don't feed	Каши не варят;
Rotten herring	На тухлых на селедках
Is all that people there can eat.	Там народ морят![45]

Centralized canteens, which replaced the cooks of workers' artels in the 1930s, did not make any difference. Complaints or jokes about their poor hygienic standards and low food quality appeared with predictable regularity on the pages of Shaturtorf's weekly newspaper *Za Torf*.[46]

Despite these privations, the transient nature of workers' settlements offered peat workers a degree of freedom, which distinguished them favorably from their home villages. Temporary labor migration had always been more than an economic strategy to increase rural households' income, constituting an important social and cultural experience for young people from the countryside. Although peat workers lamented the quality of housing and food, they seem to have cherished the opportunity to break through the patriarchal norms of rural society and to escape boredom and control at home. Folklore and oral testimonials suggest that friendship, courtship, and sexual adventure defined their experience at least as much as the hard labor and poor living conditions.

[44] Filtzer, *Soviet Workers and Late Stalinism*, 6–7, 88–99.

[45] Iskrin, *Na torfianykh bolotakh*, 10. I am grateful to Sarah Smyth for assisting me with the translation.

[46] "V stolovykh torfopredpriiatii navesti Bol'shevistskii poriadok," *Za Torf* (August 4, 1934): 1; "Pochti s natury," *Za Torf* (August 29, 1957): 2.

In a song recorded in the region of Bogorodsk, workers' settlements appear as informal spaces at the fringes of society, where peat workers seek and find romantic love:

In our bog no-one cares To sweep the barracks As the sun sets They walk around in pairs ...	На нашем на болоте Там бараков не метут; Как солнце на закате, Так по парочкам идут...
A hoarse voice can be heard implore Far and near: "Let's get together, if only for The briefest of meetings, my dear."	Голосочек хриповат Раздается там и тут: „Приди, милка, на свиданье, Хоть на десять минут!"[47]

E. A. Stanovkina, a woman from Mordovia, regularly traveled to peat extraction sites during the early 1950s. For her, the collective spending of time outside of formal working hours was a crucial part of what it meant to be a peat worker: "in the evenings, especially on free days, we had fun from the heart, singing songs in Moksha language and dancing to the balalaika."[48] Seasonal employment in peat extraction, however, could also be an abusive experience. While the marginality of peat workers' settlements opened up spaces for informal, sometimes queer sexual encounters, gendered violence, often in conjunction with excessive drinking, was regularly recorded, too. It also featured in literary depictions of these places.[49]

The relationship between peat trusts and workers changed profoundly in the second half of the twentieth century, particularly following Stalin's death in 1953. The trusts now played a fundamental role not only as producers of industrial fuel but also as providers of municipal infrastructure, which gradually began to replace or complement temporary arrangements. This development gained momentum due to the central government's housing campaign, which led to a massive increase in the housing stock, enabling millions of Soviet citizens to move into private apartments in standardized panel or brick buildings. Political leaders and

[47] Iskrin, *Na torfianykh bolotakh*, 11–12. I am grateful to Sarah Smyth for assisting me with the translation.

[48] Nikishova, *El'nikovtsy*, 662. Moksha is a Mordvinic language spoken in Mordovia.

[49] Dan Healey, "Comrades, Queers, and 'Oddballs': Sodomy, Masculinity, and Gendered Violence in Leningrad Province of the 1950s," *Journal of the History of Sexuality* 21, no. 3 (2012): 496–522; Bird, "Poetics of Peat," 597–599; Sergei Dovlatov, *The Zone*. Translated by Anne Frydman. Edited by Katherine Dovlatov (London: OneWorld Classics, 2011), 173–174.

TABLE 4.1 *Shaturtorf's housing stock, square meters. Own calculations in parentheses.*[50]

Year	Barracks	Dormitories	Houses	Total
1933	105,619	n.a.	27,619	(133,238)
1938	113,597	n.a.	46,637	(160,234)
1941	88,132	n.a.	50,956	(139,088)
1945	89,720	n.a.	60,360	(150,080)
1956	n.a.	n.a.	n.a.	186,547
1973	15,621	2,488	(210,809)	228,918
1977	3,368	3,373	(216,064)	222,805
1980	2,580	3,464	(222,458)	228,502

the media presented the new apartments as the embodiment of modern life in a socialist guise: Soviet citizens would live in nuclear families and enjoy the conveniences of consumer goods, though without succumbing to the excesses of capitalist mass consumption.[51]

The government's commitment to improving workers' daily lives coincided with a dramatic transformation within the peat industry. Mechanization drastically reduced the demand for seasonal labor. Accommodating smaller numbers of regular employees and their families around the year now became much more important than catering for many workers who stayed only for short periods of time. Between the early 1930s and the early 1980s, the living space provided by the Shatura Peat Trust grew continuously, with a notable shift away from barrack-type accommodation toward the provision of basic comfort (Table 4.1). Of the 24,346 people housed by the trust in 1980, only 173 lived in dormitories and 206 in barracks. Sixty-eight percent of Shaturtorf's residential buildings were connected to the sewerage system, 70 percent had central heating, and 61 percent were supplied with gas. However, hot water remained a luxury, being available in just 15 percent of the buildings.[52]

[50] Shatura Peat Trust, Annual Reports. 1933 (RGAĖ, f. 7969, op. 1, d. 77, l. 13ob.); 1938 (RGAĖ f. 7969, op. 1, d. 249, l. 110); 1941 (RGAĖ f. 7969, op. 2, d. 199, l. 54); 1944 (RGAĖ, f. 7969, op. 2, d. 262, l. 87); 1955 (RGAĖ f. 7979, op. 2, d. 539, l. 110ob.); 1973 (GARF f. A-632, op. 2, d. 532, l. 65); 1977 (GARF, f. A-632, op. 9, d. 148, l. 73); 1980 (GARF, f. A-632, op. 9, d. 293, l. 72). For the earlier years, the statistics did not make a distinction between communal accommodation in barracks or dormitories. The forms only list "barracks."

[51] Susan A. Reid, "Khrushchev Modern: Agency and Modernisation in the Soviet Home," *Cahiers du Monde russe* 47, no. 1–2 (2006): 227–268.

[52] Shatura Peat Trust, Annual Report (1980), GARF f. A-632, op. 9, d. 293, l. 72.

Welfare services and leisure facilities also reflected the trust's growing importance as a regional developer and the transformative impact of the peat industry on people's lives in central Russia. In 1955, all of Shaturtorf's peat companies were actively constructing medical stations, hospitals, schools, and kindergartens.[53] When the trust began to tap peat deposits located a few kilometers north-east of the city of Shatura in the late 1960s, preparations included not just the design of drainage canals and the commissioning of new extraction sites but also substantial investments into civilian infrastructure. Two workers' settlements were added to the existing settlements of Kerva and Severnaia Griva, which soon accommodated more than 5,000 people, most of them in two-story wooden or brick buildings containing 16 apartments each. The new settlements also received nurseries and kindergartens, a gym, a public laundry, a shop, a hotel, a pharmacy, and an ambulance with X-ray facilities.[54] Meeting citizens' demands remained a struggle until the very end of the Soviet period. Even so, the incorporation of welfare infrastructure and services, incomplete and rudimentary as they mostly remained, into the design of settlements marked a considerable improvement compared to the neglect of worker's most basic needs in previous decades.

Yet planning and construction by themselves did not make workers' settlements *places*. More decisive was people's active engagement with their settlements as material and social environments. In the words of the critical geographer David Harvey, "places acquire much of their permanence as well as much of their distinctive character from the collective activities of people who dwell there, who shape the land through their activities, and who build distinctive institutions, forms of organization, and social relations within, around or focused on a bounded domain."[55] The state always had a presence in this process. Following Stalin's death, the Soviet administration became highly invested in policing mundane social deviancy and teaching citizens to respect public order.[56] These policies resonated in peat workers' settlements. From the mid-1950s,

[53] Shatura Peat Trust, Annual Report (1955), RGAĖ f. 7979, op. 2, d. 539, ll. 173–174.

[54] "On the commissioning of the Petrovskoe Peat Company of the Shatura Peat Trust (17 December 1969)," GARF f. A-632, op. 2, d. 287, ll. 14–16; "Drugim stal poselok," *Za Torf* (February 26, 1982): 1.

[55] David Harvey, *Justice, Nature and the Geography of Difference* (Cambridge, MA: Blackwell Publishers, 1996), 310.

[56] Brian Lapierre, "Making Hooliganism on a Mass Scale: The Campaign against Petty Hooliganism in the Soviet Union, 1956–1964," *Cahiers du Monde russe* 47, no. 1/2 (2006): 349–375.

clubs, libraries, and stadiums allowed the workers of the Shatura Peat Trust to socialize in ways that accorded with contemporary norms of accepted social behavior. Collective events during which workers and their families gathered to tidy up or enhance public spaces were no less critical. As part of wider urban "improvement" (*blagoustroistvo*) and "greening" (*ozelenie*) campaigns, the inhabitants of Shaturtorf's settlements participated in regular community workdays (*subbotniki*), during which they planted trees, arranged flower beds, collected rubbish, or painted public facilities.[57]

Urban improvement and urban greening were ambivalent endeavors. During Stalinism, related campaigns had been closely linked to the state's repressive policies aimed at controlling people and public spaces.[58] In later periods, Soviet citizens would become frustrated by the expectation to give their free time to do "volunteer" work in their neighborhoods. Still, even though they tied workers to the state, these campaigns also instilled a sense of attachment to one's local community. According to the historian Susan E. Reid, popular mobilization to deal with quotidian neighborhood affairs provided "the cement that kept Soviet society together."[59] As the anthropologist Aleksandra Kasatkina argued, late Soviet urbanism, with its emphasis on active participation, fostered a sense of belonging and local identity among the urban population that lives on in memories decades after the Soviet collapse.[60] I felt an echo of the late Soviet approach to the making of place in the summer of 2016 when visiting Tugolesskii Bor, a peat workers' settlement near Shatura. Lashing rain prevented my colleagues and me from exploring the abandoned extraction sites whose location I had identified through satellite images. Instead, we ended up in the local House of Culture, where

57 "Zabota o blage naroda," *Za Torf* (August 29, 1957): 1; "Blagoustraivaiutsia, khorosheiut torfianye poselki," *Za Torf* (August 29, 1957): 1; "Kinofestival' v Kerve," *Za Torf* (January 25, 1964): 2; "Molodezhnoe kafe v Kerve," *Za Torf* (September 31, 1964): 2; "Torfianym poselkam khoroshet'," *Za Torf* (May 13, 1972): 2.

58 Johanna Conterio, "Controlling Land, Controlling People: Urban Greening and the Territorial Turn in Theories of Urban Planning in the Soviet Union, 1931–1932," *Journal of Urban History* 48, no. 3 (2022): 479–503.

59 Susan A. Reid, "Building Utopia in the Back Yard: Housing Administration, Participatory Government, and the Cultivation of Socialist Community," in Karl Schlögel, ed., *Mastering Russian Spaces: Raum und Raumbewältigung als Probleme der russischen Geschichte* (Munich: De Gruyter Oldenbourg, 2011), 149–185, 160. See also Bolotova and Stammler, "North," 207–209.

60 Aleksandra Kasatkina, "K poėtike i praktike modernogo urbanizma: 'Gorod-sad' kak tekhnologiia sotsial'noi integratsii v Obninske 1960-kh godov," *Laboratorium: Zhurnal sotsial'nykh issledovanii* 14, no. 1 (2022): 30–59, 41–50.

three women shared their memories from the times before peat extraction ended in the region: the school whose workshops had been built by the parents; the library (still operational when we visited), which had always been an important meeting point; and the House of the Pioneers where people had often come together. By naming these places, the women articulated a deep attachment to their settlement which for them was defined as much in social as in material terms.[61]

Growing food in private garden plots was integral to workers' place-making. Despite Stalin's violent collectivization campaign, household farming always formed a vital part of Soviet food production. It was formally recognized in the mid-1930s when the political leadership permitted subsistence agriculture to compensate for the low output of the newly created collective farms. Except for a few attempts to curtail this form of food production, subsequent governments maintained a pragmatic stance on citizens' garden plots. Producing considerable amounts of vegetables, fruit, eggs, and dairy products, the "private auxiliary farm" (*lichnoe podsobnoe khoziaistvo*) constituted an essential pillar of the food system up until the end of the Soviet period and through the 1990s.[62] Workers' settlements confirmed this pattern. Formally, they fell under the category of "urban-type settlements" (*poselki gorodskogo tipa*), whose inhabitants, by definition, pursued mainly nonagricultural work.[63] In reality, workers' settlements muddied distinctions between the rural and the urban. While the issue of food production in such settlements had been contested in the early Soviet period, as some saw it as conflicting with revolutionary ideals, household farming was soon considered an intrinsic feature of Soviet workers' settlements.[64] According to a 1940 manual for planners, sufficient land was to be allocated to gardens and pasture so that between 25 and 30 percent of the local demand for

[61] Collective interview with the author, Tugolesskii Bor, July 7, 2016.

[62] Andrei Markevich, "Finding Additional Income: Subsidiary Agriculture of Soviet Urban Households, 1941–1964," in Donald Filtzer, Wendy Z. Goldman, Gijs Kessler, and Simon Pirani, eds., *A Dream Deferred: New Studies in Russian and Soviet Labour History* (Bern: Peter Lang Press, 2008), 385–415; Stefan Hedlund, *Private Agriculture in the Soviet Union* (London: Routledge, 1989); Susanne A. Wengle, *Black Earth, White Bread: A Technopolitical History of Russian Agriculture and Food* (Madison, WI: University of Wisconsin Press, 2022), 104–111.

[63] The number and population of urban-type settlements grew considerably over the Soviet period. In 1989, 24.1 million Soviet citizens, or just over 12 percent of the urban population, resided in settlements of this category. Goskomstat, *Naselenie SSSR 1988: Statisticheskii ezhegodnik* (Moscow: Finansy i statistika, 1989), 4.

[64] On these early Soviet debates, see Meerovich, *Gradostroitel'naia politika*, 280–284.

potatoes and other vegetables could be satisfied.[65] Throughout the Soviet Union, most people living in workers' settlements were involved in producing food. Being no exception, peat workers spent quite a considerable part of their time outside of formal working hours growing vegetables or caring for chickens or cows.

Gardening was never a fully autonomous undertaking though. Given that the produce from workers' plots was vital for keeping the Soviet population fed, state organs actively purchased privately produced food. They also assigned food production targets to industrial enterprises, which in turn passed these down to their employees. For peat workers, the connection between their own produce and state norms could be both a source of frustration and an opportunity to leverage their personal interests. Take the complaint submitted to the Presidium of the Supreme Council by the trade union of Shaturtorf's company Radovitskii Mokh in 1953. As the union pointed out, two of the places where the company's peat workers lived were not formally recognized as workers' settlements and thus lacked some of the basic infrastructure associated with this status. The unjust nature of the arrangement was at the heart of the letter. While the authorities registered the land, livestock, and population of these localities, the union complained, their residents received very little in return. Since workers' homes were located several kilometers away from the next rural council (*Sel'sovet*), solving even the most trivial bureaucratic matter was extremely time-consuming. The fact that neither place had any milk collection points (*slivnye punkty*) added to their hardships, as workers had to walk long distances to deliver their home-produced milk. Pointing to the upcoming expansion of the Radovitskii Mokh peat company, which was expected to attract more people to the area, the petition asked that both places be formally recognized as workers' settlements. In early 1954, the regional administration satisfied this request at least for Radovitskii Mokh, a seemingly trivial change, but certainly an improvement as it meant that workers could expect better services and facilities.[66] As time went on, food production in workers' settlements became more, not less important. From the early 1970s, when the Soviet government began actively promoting private household farming,

[65] Ia. P. Levchenko, *Planirovka rabochikh poselkov* (Moscow: Izdatel'stvo NarkomKhoza RSFSR, 1940), 16.

[66] "Ob otnesenii naselennogo punkta Radovitskii Mokh Klepikovskogo raiona Riazanskoi oblasti k kategorii rabochikh poselkov (1954)," GARF f. A-385, op. 17, d. 2584, ll. 11, 13, 21–22.

Shaturtorf's administration regularly appealed to its employees, calling them to produce more food, especially dairy. The Shatura Peat Trust also extended its network of delivery points, and the trust's newspaper published gardening advice and cheered communities, which over-fulfilled the delivery quota.[67]

Household farming gave workers reliable access to fresh food, making them autonomous from poorly supplied shops and canteens. But it also served an important emotional role, eliciting feelings of purpose and belonging. Late Soviet sociologists sometimes pointed to this intangible dimension of household farming to justify and demand state support for private plots.[68] In his research on working-class communities in contemporary Russia, anthropologist Jeremy Morris identified self-provisioning among the many "fragile, unfinished and improvised forms and practices of life-making in the Russian margins," functioning as both coping mechanisms and sources of meaning that together make the world "habitable."[69] The connection between personal agency, habitability, and attachment to place resonated strongly during my conversation with local women in Tugolesskii Bor in 2016. The pivotal role of garden plots back in the days when the settlement's economy revolved around peat extraction was without question for them: "Everyone had their plot (*svoi nadel*)," one commented. Looking back, her friend cast private food production as a source of plenty and reliability: "We planted vegetables: cabbage, cucumbers, tomatoes – everything was there and grew." That local workers produced potatoes and squash and fermented their own cabbage was also mentioned with pride.[70]

Caring for animals, digging up potatoes and fodder beets, or haymaking even became part of peat workers' official work life. During the Brezhnev period, when Soviet agriculture began facing chronic labor shortages, the relationship between peat extraction and the farming sector changed profoundly. While most workers had come from an agricultural background and returned to the villages in the early phases of industrial peat extraction, workers' mobilities now took the opposite direction, as peat-extracting enterprises routinely delegated their employees to help out

[67] "Tsennaia initsiativa," *Za Torf* (March 4, 1972): 2; "Sovety ovoshchevodam," *Za Torf* (September 9, 1972): 2.

[68] Katja Bruisch, "The Soviet Village Revisited: Household Farming and the Changing Image of Socialism in the Late Soviet Period," *Cahiers du Monde russe* 57, no. 1 (2016): 81–200, 91–93.

[69] Morris, *Everyday Post-Socialism*, 9, 34, 37.

[70] Collective interview with the author, Tugolesskii Bor, July 7, 2016.

on state farms.[71] Peat trusts also set up their own auxiliary farms (*podsobnye khoziaistva*), which were a fairly common feature of industrial enterprises, particularly after the adoption of the Food Program in 1982.[72] As a result, pig pens and storehouses for vegetables and other agricultural produce became part of the production infrastructure of Shaturtorf's companies. Alongside quotas for peat products, the peat industry now also received annual targets for agricultural goods, which individual enterprises were expected to produce on their own premises. Yet, amidst the country's deepening economic crisis, peat companies were no better positioned than most state farms to make food production more efficient. In 1987, most peat trusts in the RSFSR did not meet their agricultural quotas. With just 45.5 percent plan fulfilment, Shaturtorf's performance was particularly weak.[73]

These histories hint at the in-betweenness of peat workers' settlements in the social and economic fabric of the Soviet state. Peat workers were not only providers of fuel and other peat-based products but also performed essential functions of social reproduction as they labored in fields, orchards, gardens, and sheds to provide food for themselves and the state. In the concrete environments of their settlements, it was impossible to delineate these different economic roles from one another. Labor for private needs and labor in formalized relationships constantly overlapped. In the late 1980s, one of Shaturtorf's companies regularly asked its workers to fatten its pigs at home because the company-owned pigpen was overcrowded.[74] For the women I met in Tugolesskii Bor, the line between food production for private and collective needs was blurry, too. As we talked, it emerged that the settlement had once had a clearly defined

71 *Za Torf* regularly reported about peat workers' contribution to food production on state farms. "Pod-em sel'skogo khoziaistva - vsenarodnoe delo," *Za Torf* (January 25, 1964): 1; "Na sovkhoznykh poliakh," *Za Torf* (September 27, 1972): 1; "Po dolgu shefov," *Za Torf* (May 28, 1983): 2; Shatura Peat Trust, Annual Report (1978), GARF A-632, op. 9, d. 198, l. 130.

72 Laure Després and Ksenya Khinchuk, "The Hidden Sector in Soviet Agriculture: A Study of the Military Sovkhozy and Auxiliary Farms," *Soviet Studies* 42, no. 2 (1990): 269–293; Wengle, *Black Earth, White Bread*, 108.

73 "Prikazy Rostorfa za 1981 god," GARF f. A-632, op. 9, d. 298, l. 173; "O neudovletvoritel'noi rabote podsobnykh sel'skikh khoziaistv proizvodsvennykh ob-edinenii po itogam I-go polugodiia 1987 goda," GARF f. A-632, op. 9, d. 478, ll. 43–45; "Ob utverzhdenii gosudarstvennoi komissii pro priemke na ėkspluatatsiiu svinarnika na 100 golov na Petrovskom torfopredpriiatii Shaturskogo proizvodstvennogo ob-edinenija po dobyche torfa (1984)," GARF f. A-632, op. 9, d. 400, l. 160; "Dve fermy - dva rezul'tata," *Za Torf* (March 2, 1990): 1.

74 "I miaso i kolbasy," *Za Torf* (April 3, 1990): 1.

geography of animal husbandry, where cows were not housed centrally, yet were, at the same time, considered as belonging to a larger herd (*stado*): "Look, our street and the workers' settlement (located a bit outside of the main settlement – KB), that was one herd. The Tsentralnaya, Gorky and Sovetskaya Streets – that was the second herd. The October and Sovetskaya Streets – another one. Pervomay had its own. There were ninety cows in each herd."[75] The everyday in peat workers' settlements defied neat distinctions between rural and urban ways of life, between work and leisure, and between productive and reproductive labor.[76] The permeability of these boundaries allowed the Soviet economy to function in its messy and often unplanned ways while at the same time providing peat workers with the security and somewhat limited autonomy to make a home at the margins of Russia's fossil economy.

PLACE BETWEEN PATRIOTISM AND DEEP TIME

Alongside material changes in the landscape and community-building practices, myths and stories played a crucial role in the history of place-making, weaving the Meshchera Lowlands into various incarnations of Soviet patriotism. Many of these initially centered on Shatura and its famous power station. Although the idea to build such a utility in Shatura had originated in the late imperial period, construction began only a few months after the Bolshevik takeover and gained traction following the release of the GOĖLRO-plan. The first regional power plant to be commissioned after the Bolshevik takeover, Shatura was often presented as proof of the progressive nature of the Soviet state. The station was named after Lenin to mark its significance and has since been deeply ingrained in the Soviet electrification myth.[77] In 1926, the revolutionary Lev Trotsky referred to it as a beacon of modernity in a country whose culture was backward and nature untamed:

> Moscow is the capital of the Communist International. But you go a few dozen kilometres and you are in the backwoods, with snow-laden fir trees, frozen marshes, and wild beasts. Dark hamlets of log huts are dozing in the snow.

[75] Conversation with the author, Tugolesskii Bor, July 7, 2016.

[76] This was a crucial feature of late Soviet society as demonstrated by Alexandra Oberländer, "Cushy Work, Backbreaking Leisure: Late Soviet Work Ethics Reconsidered," *Kritika: Explorations in Russian and Eurasian History* 18, no. 3 (2017): 569–590.

[77] A. M. Zaikov, *Geroicheskaia Shatura: Pervenets sovetskoi ėlektrifikatsii* (Moscow, Leningrad: Gosudarstvennoe izdatel'stvo, 1928).

From the carriage window you can sometimes see the tracks of wolves. Where the Shatura station stands today, a few years ago, when they began construction work there, elks had their homes. Today the distance between Moscow and Shatura is covered by an elegant series of metal masts which carry cables with a current of 115,000 volts.[78]

The Soviet government actively sustained the Shatura myth both domestically and abroad. After ambassadors and journalists from several countries had attended the utility's opening ceremony in 1925, Shatura became a destination for foreign energy experts. In 1935, a delegation sent by the Irish Ministry for Industry and Commerce visited the plant as part of a trip to Germany and Soviet Russia, where the group studied the large-scale extraction of peat and its potential for electricity generation. In Shatura, the visitors noted not only the massive size of the extraction fields with their 32,000 workers. They were also impressed by the station's exceptional capacity. As they were shown around, one of the plant's chief engineers boasted that the efficiency of power stations like Shatura did not fall short of coal-fired utilities. Interested in exploring ways to reduce Ireland's dependence on British coal and to mobilize what some at the time called Ireland's "indigenous" resources, the visitors were pleased to hear this. Their report mentioned that power plants using peat could run efficiently, a conclusion that would inform the later decision to rely on peat for Ireland's electrification campaign.[79] The notion of Shatura as a manifestation of socialist progress lived on for decades after the plant's opening. Inside the Soviet Union, numerous official publications reiterated its central place in popular electrification narratives, while two Soviet cargo vessels, Shaturstroi (built 1918) and Shatura (built 1966), carried its name across the oceans of the world.[80]

[78] Leon Trotsky, "Culture and Socialism," in *Problems of Everyday Life and Other Writings on Culture & Science* (New York: Pathfinder Press, 4th printing 1986), 240.

[79] "Report of Peat Delegation to Germany and Russia, 17th May, 1935–2nd July, 1935," ESB Archives, Dublin; Toomey, "'A National Resource of Primary Importance'," 170–172.

[80] *Shatura: Sbornik statei k 25 letiiu GRĖS imeni V. I. Lenina* (Moscow, 1951); A. V. Vinter, "Ot Shatury do DneproGĖSa," in *Sdelaem Rossiiu ėlektricheskoi: Sbornik vospominanii uchastnikov Komissii GOĖLRO i stroitelei pervykh ėlektrostantsii* (Moscow: Gosudarstvennoe ėnergeticheskoe izdatel'stvo, 1961), 32–38; G. Lipenskii, *Bol'shaia Shatura* (Moscow: Moskovskii Rabochii, 1970); on the vessels see *Various Data on the Merchant Marine Fleets of the United States and Other Major Sea Powers* (Washington: US Government Printing Office, 1970), 11755–11756; www.balticshipping.com/vessel/imo/6700078. Accessed March 25, 2025.

In contrast to the Shatura myth, whose contours stabilized quite early in the Soviet period, the meanings invested in the town's surroundings changed considerably over time. Cultural representations of peat extraction sites and workers' settlements reflected Soviet visions of cultural transformation, particularly the aspiration of turning backward peasants into conscious workers and builders of socialism. Looking back at his time as an administrator in Shatura, a certain Iazykov described how, in the early 1920s, peat workers used to pray at the beginning of the season and brought icons and a priest to the extraction sites. In one instance, workers even established a power connection from the provisional Shatura station to illuminate a local church for the upcoming Easter holidays.[81] Echoing Soviet elites' disregard for rural and religious culture, Iazykov's account marked the places where peat workers lived and worked as starting points of their emancipatory journey – a view that featured prominently in other sources too. As the women's journal *Rabotnitsa* put it in 1928, it was while being peat workers that thousands of women from the countryside "acquire first cultural skills, receive knowledge, learn to see life with different eyes."[82] Such ideas structured time and space at peat extraction sites and adjacent settlements, as peat-extracting companies organized reading circles, clubs, public lectures, and "red corners." Even though often underfunded, these efforts were presented as tools to make peat workers ambassadors of the Soviet project. From the peat works, the new culture would radiate into the darkest corners of the Soviet state: "The *torfushki*, having learned to read and write, will take their knowledge to the village."[83] During the late Soviet period, the emphasis of the discourse shifted from transition to achievement, mission to consolidation, revolution to normality. In 1964, Shaturtorf's newspaper *Za Torf* celebrated peat workers' accomplished transformation: "The modern (*sovremennyi*) peat worker – this is not the 'man with the spade' anymore, who was illiterate and driven into the swamp by a bitter, desperate fate. The modern peat worker is an industrial worker, who is in control of the technology, who uses difficult machines."[84] The worker evoked in these lines was skilled and undeniably male – a stark contrast to the illiterate woman with a headscarf that had dominated earlier representations of the peat industry.

81 "Vospominaniia o Shature," GARF f. 7952, op. 3, d. 459, ll. 74–75.
82 "Na torforazrabotkakh," *Rabotnitsa* no. 23 (June 1928), 5.
83 "Za odnogo gramotnogo – dvukh negramotnykh," *Rabotnitsa* no. 28 (July 1928), 13.
84 "Chelovek s lopatoi ushel v proshloe," *Za Torf* (January 25, 1964): 2.

Complementing the tale of workers' cultural accomplishment was the notion of peat workers' settlements as quintessential Soviet places, invoked through the reproduction and appropriation of the official discourse in the spaces of the everyday. The streets and squares of peat workers' settlements served as sites for annual parades commemorating the October Revolution or Victory Day, as clubs, schools, and Houses of the Pioneers hosted lectures and public meetings or put up posters to mark important events in the Soviet calendar. The anthropologist Aleksei Yurchak pointed to the importance of the past for late Soviet discourse and the strong performative dimension of propaganda and rituals, whose actual meaning became "open-ended, indeterminate, or simply irrelevant."[85] Crucially for our case, the conventions of the authoritative discourse framed local tales of origins in peat-extracting regions that gained shape in the post-Stalinist period. Lenin was central to this new idea of place that hinged on both formulaic readings of the Soviet past and people's personal links to the peat industry. In the spring of 1960, when the country celebrated 90 years since Lenin's birth, Shaturtorf's newspaper *Za Torf* did not tire of stressing how crucial Lenin's support for the peat industry had been for the region. Revisiting the transformation that Shatura and his home settlement Shaturtorf had undergone in recent decades, the pensioner M. Chudin enthused: "Lenin's dream has come true."[86] The Lenin cult also translated into the built environment, as Lenin Streets and monuments of the revolutionary leader underscored the rootedness of workers' settlements within the broader trajectory of the Soviet project (Figure 4.4).

Some stories evoked a sense of place by venturing much further back in time. These focused not on Shatura as a symbol of Soviet electrification but on the Meshchera Lowland as a distinct geographical and cultural space. The region had become a point of reference in the late 1930s, when the writer Konstantin Paustovskii published his tale "Meshchera land" (*Meshcherskaia storona*) celebrating the harmony of Russian nature and culture in Meshchera's ancient history.[87] During the late Soviet period,

85 Aleksei Yurchak, *Everything was Forever, Until It was No More: The Last Soviet Generation* (Princeton: Princeton University Press, 2006), 26.

86 "Lektsii i besedy ob Il'iche," *Za Torf* (March 23, 1960): 1; "Po puti, ukazannomu Leninym, vpered, k kommunizmu," *Za Torf* (April 22, 1960): 1; "Predvidenie vozhdia," *Za Torf* (April 22, 1960): 2; "Oni byli pervymi," *Za Torf* (April 22, 1960): 2; "Tak stroilas' Shaturskaia GRĖS," *Za Torf* (April 22, 1960): 2.

87 Laurent Coumel, "La Mechtchera au XXe siècle: L'environnement russe entre ressource(s) et patrimoine," *Siècles* no. 42 (2016), https://doi.org/10.4000/siecles.2962.

FIGURE 4.4 Lenin monument in Tugolesskii Bor. *Source:* Author's photograph (2016).

interest in the region's more distant past was closely tied to the revival of *kraevedenie*, a discipline and movement devoted to studying local history, customs, and nature whose origins dated back to the imperial period. Museums, exhibitions, and domestic tourism now began to integrate the imperial past and the achievements of Soviet socialism into a linear historical narrative. *Kraevedenie* was a tool through which the post-Stalinist leadership promoted place-based identity as an articulation of Soviet patriotism, which increasingly incorporated motives of ethnocentric Russian nationalism. Actively encouraged from above, *kraevedenie* also opened up new spaces for local initiatives to preserve historical architecture and other monuments.[88]

Late Soviet Meshchera *kraevedenie* fused official discourse and bottom-up initiative. One local enthusiast was N. N. Akimov, a schoolteacher from Pustoshi near Shatura, amateur archaeologist, and head of the Meshchera Historical Section, who made several remarkable discoveries of Neolithic ceramic fragments and flint tools at Shatura's Sviatoe lake. In the 1970s, when Shaturtorf's newspaper *Za Torf* began to

[88] On local tourism and the intersection of Soviet patriotism and Russian nationalism in the late Soviet period, see Victoria Donovan, *Chronicles in Stone: Preservation, Patriotism, and Identity in Northwest Russia* (Ithaca: Northern Illinois University Press, 2019), 75–83.

dedicate some space to local studies, Akimov regularly shared his findings.[89] Many articles in the new rubric, titled "My peaty land" (*Torfianaia moia storona*) in a nod to Paustovskii, stressed that Meshchera's history encompassed much more than the peat industry, with which readers typically associated it. In 1972, *Za Torf* reported about the Soviet government's decision to allocate state funds to preserve the Nikolo-Radovitsky Monastery, whose major buildings dated from the seventeenth century. A photograph above the short note centered on one of the monastery's onion-domed churches, a signifier of Russia's cultural heritage at the time. Several weeks later, an article by Akimov placed the monastery and its surroundings in a historical tradition going back thousands of years. Mentioning archaeological evidence of "primitive" Stone Age settlement, the text moved through the period of medieval "old Russian" culture, the "Tatar invasion" of "Russian land" (*russkaia zemlia*), and the ensuing revival of Russian culture evidenced by the influence and wealth of the Nikolo-Radovitsky Monastery. The article culminated in a celebration of Soviet power and statehood, which had "woken up the age-old silence of the swampy region" and brought convenient settlements and progressive industry.[90] In its ignorance of Meshchera's multiethnic past, Akimov's tale followed the conventions of Soviet historiography, which emphasized the resilience and superiority of Russian culture while depicting the Tatars in overtly pejorative terms.[91] On the pages of *Za Torf*, such simplified narratives constructed place not only in spatial but also in temporal terms, conjuring a history so rich that it overshone the social dynamism and ecological ruptures that had massively transformed the region since large parts of its nature had become subject to the imperative of extraction.

It was due to the specific nature of peatlands that conceptions of place in Meshchera also incorporated deeper levels of time. The waterlogged conditions, which inhibit the decomposition of organic carbon-rich matter, also make these ecosystems unique archives of the distant human past. Within a peatland, pollen, remains of animals or plants, as well as

[89] V. V. Sidorov, "Stoianki na ozere Sviatom u Shatury," *Sovetskaia Arkheologiia* no. 3 (1975), 107–117.

[90] "(without title)," *Za Torf* (January 27, 1972): 4; N. Akimov, "Radovitskaia mshara," *Za Torf* (March 15, 1972): 4.

[91] Bulat R. Rakhimzianov, "Meshchera as a Point of Political Interaction between Muscovy and the Tatar World," *Russian History* 43, no. 3–4 (2016): 373–393; Charles J. Halperin, "Soviet Historiography on Russia and the Mongols," *The Russian Review* 41, no. 3 (1982), 306–322.

human bodies, tools, and other cultural artefacts remain intact for millennia as long as the water table is high enough to shield them from exposure to oxygen.[92] The significance of peatlands as historical record keepers typically becomes apparent when they get destroyed, a pattern also evident in the history of the Soviet peat industry. In 1969, ancient human artefacts were unearthed during drainage and extraction works near Tugolesskii Bor. Initially confined to a local newspaper, news of the find prompted people from the area to explore the site on their initiative. Although local enthusiasm was crucial in drawing the attention of professional archaeologists, members from the Academy of Sciences, who arrived only after several years, lamented later that the amateurs had damaged archaeological evidence using bulldozers and other heavy machinery for their explorations. Known as *Voimezhnoe I*, the site is nowadays recognized as a monument of Stone and Bronze Age settlement situated within a larger landscape shaped by prehistoric settlement.[93] Both within the region and beyond, archaeological discoveries made during the advance of the peat industry through central Russia's wetlands shaped how Meshchera's placeness has been constructed. After some items were displayed in a small informal exhibition in Shatura, most artefacts from Tugolesskii Bor ended up in the Moscow Regional Museum of Local Lore. When Shatura received a Museum of Local Lore in 1987, its permanent exhibition included pre-historic objects found during several excavations undertaken by the Academy of Sciences in the region.[94] Deep Time had now become a prominent feature of Shatura's official history.

* * *

[92] Benjamin Gearey and Ralph Fyfe, "Peatlands as Knowledge Archives," in Aletta Bonn, Tim Allot, Martin Evans, Hans Joosten, and Rob Stoneman, eds., *Peatland Restoration and Ecosystem Services: Science, Policy and Practice* (Cambridge: Cambridge University Press, 2016), 95–113.

[93] On the history and significance of the archaeological research near Tugolesskii Bor, see A. S. Aleshinskaia et al., *Drevnie okhotniki i rybolovy Podmoskov'ia po materialam mnogosloinogo poseleniia ėpokhi kamnia i bronzy – Voimezhnoe I* (Moscow: RAN Institut Arkheologii 1997), 4–7; V. V. Dvornichenko and V. V. Sidorov, "Novaia stoianka na tugolesskikh ozerakh," *Arkheologicheskie Otkrytiia* (1973): 62–63; Mikhail G. Zhilin, "Early Mesolithic Hunting and Fishing Activities in Central Russia: A Review of the Faunal and Artefactual Evidence from Wetland Sites," *Journal of Wetland Archaeology* 14, no. 1 (2014): 91–105. Shaturtorf's newspaper *Za Torf* regularly reported about the area's pre-historic heritage. N. Akimov, "Tugolesskii Bor," *Za Torf* (January 29, 1972): 4; A. L'vov, "Na beregu ozera Karasovo," *Za Torf* (October 18, 1972): 4.

[94] "Vystavka drevnostei," *Za Torf* (January 29, 1972): 4; https://shaturamuseum.ru/o-dk/istoriya/.

A century of industrial peat extraction left Meshchera's wetlands profoundly transformed. In contrast to the late imperial sources, which had evoked the image of a poorly integrated and thinly settled space with abounding wildlife, the region turned into an industrial hinterland once its wetlands began to supply ever-growing amounts of fuel, which helped to produce electric power for surrounding enterprises and the city of Moscow. The conversion of peatlands into internal resource frontiers of Russia's expanding fossil economy meant that human and nonhuman life in the Meshchera increasingly revolved around the rhythms and needs of extraction. Shatura, which had been a settlement of just around 5,000 inhabitants in the 1920s, had grown into a town of nearly 30,000 in 1979. In the same year, the Shatura Peat Trust housed over 23,000 people, most of whom lived in urban-type settlements in the town's vicinities.[95] Surrounding these settlements was a landscape defined by rectangular extraction fields, drainage ditches and canals, railway tracks, and roads. The spatial orders these technologies imposed on central Russian wetland environments reflected the priorities of an energy-hungry economy and the power imbalances between social elites and the wider population. Their imprint on the land was so deep that extraction has retained a physical presence long after it ceased.

For those employed in the peat industry, Meshchera's transformation was an ambivalent experience that involved departing and arriving, extracting from and connecting with the land, as well as multiple encounters with larger politics of marginalization and inclusion. The evolution of peat extraction from an activity dominated by seasonal manual labor toward a more regular industry involved corresponding shifts in the demography and characteristics of the workforce. From a bird's eye perspective, it may seem that these advanced along a linear trajectory of progress – from peasant to industrial, from unskilled to skilled, from female to male, from traditional to modern. Contemporary commentators liked to dwell on Soviet workers' successful metamorphoses, but the reality on the ground was much messier. Seasonality never lost its role as a structuring factor of the peat industry, while workers, even once they began to settle in peat-extracting regions permanently, always remained

[95] www.shatura.ru/about/history?ysclid=leycc2ee05101133 45. Accessed March 7, 2023; "Vsesoiuznaia perepis' naseleniia 1979 g." https://web.archive.org/web/20190426073119/http://www.demoscope.ru/weekly/ssp/rus79_reg2.php. Accessed March 7, 2023; "Annual report, Shatura Peat Trust (1979)," GARF f. A-632, op. 9, d. 247, l. 68.

engaged in primary food production. At the margins of Russia's fossil economy, industrial and agricultural labor, production and reproduction always remained closely intertwined.

Still, the late Soviet period brought qualitative changes at the societal level as much as in the lived experiences of peat workers. While, up until the 1950s, most settlements housing the industry's workforce had been transitional spaces to which few people felt a personal attachment, they began to reflect the state's growing commitment to care for the population's basic needs and the idea of "normality" which was paramount in late Soviet discourse. When compared to industrial agglomerations such as the famous city of Magnitogorsk, small company towns in the Soviet Union were disadvantaged in terms of budget allocation and social welfare.[96] Most of Shaturtorf's workers could not count on the conveniences of daily life facilitated by increasing energy consumption and civil infrastructure that gradually became more common in large cities. They often lacked sufficient transport facilities, lived in poorly maintained houses, and had little access to well-stocked shops. Nevertheless, peat-extracting enterprises and affiliated workers' settlements became crucial units of social organization and anchors of personal belonging in the late Soviet period. Workers added to the placeness of their settlements through activities such as private food production or cultural events, as local histories provided them with a sense of rootedness in space and time. The interplay of these factors allowed workers to develop a sense of home at the margins of Russia's fossil economy and explains why the end of extraction turned into a deeply unsettling experience of loss and precarity for many of them.

[96] Mark B. Smith, "Faded Red Paradise: Welfare and the Soviet City after 1953," *Contemporary European History* 24, no. 4 (2015): 597–615, 602–603.

PART III

UNSETTLING ENVIRONMENTS: THREAT, LOSS, AND PRECARITY IN RUSSIA'S PEATLANDS

5

Irritation

That they are highly prone to catching fire was one of the first things Mikhail Prishvin understood about peat extraction sites. In the 1920s, Prishvin, a well-known writer since the prerevolutionary period, visited the region of Yaroslavl, northeast of the city of Moscow. Dispatched by the Workers' Newspaper (*Rabochaia Gazeta*), he was to report on the area's peat works just a few years after the Bolsheviks had decided to make local fuels a pillar of the Soviet energy system. Published in 1926, Prishvin's essay "Peat" (*Torf*) is a powerful testimony to the ambivalent role of the peat industry in the Soviet state, where it fueled the recovering economy while simultaneously creating multiple environmental hazards. Upon arriving at the Berendeevskoe peat company, the writer was initially impressed to see how central Russian wetlands were being transformed. Observing the peat works from an elevated piece of land, he admired how extraction blended into the surrounding landscape: "Over there, in the middle of a clearing, there were machines, looking small from here, like cranes, when they go out to roam the swamps in the early morning." But admiration quickly turned to alarm as smoke began to rise from the peatland: "I had to run, along a cranberry path, following a worker with a fire extinguisher. It was sticky and hot. The sun was beating down from above, and the swamp was steaming from below."[1]

Potential and hazard were closely intertwined in the physical spaces of the peat industry. In Prishvin's essay, the fire ultimately becomes a moment of triumph. The workers succeed in controlling it before any severe damage

[1] M. M. Prishvin, "Torf," in: *Sobranie sochenenii a, Vol. 3: Proizvedeniia 1924–1935 godov* (Moscow: Khudozhestvennaia Literatura, 1983), 489.

occurs. For the writer, the fire was an intense sensory experience. Through the thick smoke, the sun looked "as if shining through sooty glass." But Prishvin was keen to emphasize that the significance of the event extended beyond his personal experience. Given that the energy stored in peat becomes unusable for humans when a peatland starts burning, the incident was also one of material loss. Still, Prishvin stressed that the promise lying within these environments was irresistible. Peatlands carried the seeds of a future whose abundance was hard even to put in words: "Such riches are sleeping in these impenetrable swamps of ours. What will it be like in the country, once the organized labour of man wakes up this sleeping beauty?"[2]

Prishvin's essay is an illuminating example of the intricate relationship between ideas and matter in the history of peat extraction. Modern conceptions of wetlands combined aesthetic conventions, economic interest, and visions of control and order, depicting the ideal state of wetlands as one in which water would no longer impede progressive uses. Describing peatlands as "impenetrable," "wild," and "tiringly monotonous," the writer subscribed to a century-old intellectual tradition which juxtaposed the nature of peatlands with modern human ambition. Inspired by the transnational improvement discourse, bureaucrats, intellectuals, and scientists in Russia's empires called for wetlands to be drained and cultivated to foster economic development and to "civilize" the people living in regions with extensive wetland areas.[3] Prishvin's writing was more ambivalent than the improvement discourse though. Having commented on the advances of the early Soviet peat industry, the writer acknowledged peatlands' vitality and their ability to undermine human efforts designed to dominate them. His essay foregrounds the agency of fire, a problem which would present an ongoing challenge as industrial peat extraction expanded in the coming decades. Some of Prishvin's later works similarly engaged the "tension between imagination and the resistance of material," as the literary scholar Robert Bird put it.[4] Prishvin was not the only author at the time to ponder encounters between human vision and recalcitrant matter.[5] Still, his way of combining sentimental

[2] Ibid., 490.

[3] On the improvement discourse in the imperial and Soviet periods, see Katja Bruisch, "Nature Mistaken: Resource-Making, Emotions and the Transformation of Peatlands in the Russian Empire and the Soviet Union," *Environment and History* 26, no. 3 (2020): 359–382, 371–376.

[4] Bird, "Poetics of Peat," 612.

[5] The tension between ideas and resistant matter in imperial Russian and Soviet literature is discussed in Erley, *On Russian Soil.*

depictions of landscapes with a strong belief in nature improvement has puzzled contemporary and later critics, earning him a place "among the least understood Russian authors of the twentieth century."[6]

Extraction implicated humans and peatlands in a state of mutual irritation. Prishvin may not have fit the literary canon of his time, but he had a point. The tension in his work, between the ambition to render peatlands useful and the unintended consequences of such endeavors, ran like a red thread through the material history of peat in Russia's fossil economy. Peatlands are complex ecosystems in which water, plants, and microbes are involved in constant exchanges of energy and matter, leading to the accumulation of peat. When drainage and extraction disrupt these processes, peatlands may respond unpredictably, forcing humans to adapt to their continuously changing surroundings. Contemporaries learned quickly that the interplay of nonliving matter, technology, and human bodies at peat extraction sites created hazards that could be difficult to control. In its early stages, peat extraction relied on thousands of seasonal workers and involved the creation of sometimes quite extensive bodies of shallow water. Together, these factors fostered the spread of malaria, which became a massive problem in central Russia's wetlands during the 1920s. Fire was another persistent threat wherever humans lowered the water table of peatlands. It became a significant concern during the second half of the twentieth century when Soviet peat trusts switched to milled peat extraction, which required deeper, more intrusive drainage and made the peat more prone to catching and accelerating fire than the sods produced in earlier times.

These irritations form part of a wider history of extractivism, Prometheanism, and ecological stress in the Russian Empire and the Soviet Union. Eurasian environments have witnessed multiple cases in which modernist projects undermined the vitality of entire ecosystems and, at times, even the goals of those same projects.[7] In comparison to

[6] Alexandra Smith, "The Enigma of Mikhail Prishvin: Prishvin's Pre-Soviet and Soviet Diaries," *New Zealand Slavonic Journal* (2000): 79–92, 79.

[7] The point has been made by Murray Feshbach and Alfred Friendly, *Ecocide in the USSR: Health and Nature under Siege* (London: Aurum Press, 1992). See as well, Douglas R. Weiner, "The Predatory Tribute-Taking State: A Framework for Understanding Russian Environmental History," in Edmund Burke III and Kenneth Pomeranz, eds., *The Environment and World History* (Berkeley: University of California Press, 2009), 276–315. For historical studies on resource depletion and ecosystem decline see, for instance, Brown, *Manual for Survival*; Demuth, *Floating Coast*; Ryan Tucker Jones, *Red Leviathan: The Secret History of Soviet Whaling* (Chicago: The University of Chicago Press, 2022); Peterson, *Pipe Dreams*.

well-known examples such as heavy metal pollution, the depletion of marine animal populations, soil salinization, nuclear fallout, or the desiccation of the Aral Sea, the ecological impact of peat extraction and the implications for social groups whose lives were tied to the peat industry seem relatively small. Rarely did these issues captivate the attention of a wider public, while historians have never paid attention to them. Still, as this chapter will demonstrate, the case of peat provides important clues about the relationship between environmental change, disaster, and vulnerability in imperial and Soviet Russia.[8] Revealing themselves in those moments when humans actively interfered with peatland ecosystems, the mundane hazards at and around peat extraction sites demonstrate the extent to which people and environments at the margins of Russia's fossil economy were exposed to risks created by material demands and desires formulated elsewhere. Affecting workers much more than planners or energy consumers, these risks marked peat extraction sites as spaces of structural environmental injustice.[9] At the same time, they reveal the dialectics between wetlands and wastelands that accompanied the appropriation and exploitation of peatlands. As in other countries, social elites in Russia had long resented peatlands as unproductive and treacherous landscapes that ought to be drained and used in the name of progress. Yet, as the history of malaria and fires at peat extraction sites discussed in this chapter suggests, it was the attempt to tame these landscapes that made irritation, and sometimes confrontation, a persistent thread in human–peatland interactions.

BREEDING DISEASE

Among the many reasons for the widespread animosity against wetlands, which went back to the Enlightenment period, disease had always been crucial. Scientists in the Russian Empire noted early on that people in regions where wetlands formed a prominent part of the environment tended to suffer from poor health. Fevers in particular seemed to prove

[8] Andy Bruno, "Tumbling Snow: Vulnerability to Avalanches in the Soviet North," *Environmental History* 18, no. 4 (2013): 683–709. For an introduction to disaster studies and social vulnerability in Eastern Europe, see Marc Elie and Klaus Gestwa, "Zwischen Risikogesellschaft und Katastrophenkulturen: Zur Einführung in die Katastrophengeschichte des östlichen Europas," *Jahrbücher für Geschichte Osteuropas* 62, no. 2 (2014): 161–179.

[9] On the political ecology of the workplace, see Barca, "Labouring the Earth," 11–16 and Schaupp, *Stoffwechselpolitik*.

the noxious quality of such landscapes. Underpinning this idea was the internationally accepted miasma theory, which assumed that, in marshy areas, particles of decaying material entered the air and penetrated the human body to cause disease. The primary framework to explain malaria until it was identified as a vector-borne disease, miasma theory informed the thinking of improvers in imperial Russia. As one author put it in 1796, "all swamps, whatever their kind, always have a more or less rotten vapour which spoils the air," and negatively affected human health.[10] Hints at wetlands' "rotten vapors" (*gnilye ispareniia*) were common until the late imperial period.[11] The presumed link between people's health and the landscapes they inhabited fit squarely with the idea that wetlands represented nature's faultiness, which humans had to undo. The entry "Swamp fever (malaria)" (*bolotnaia likhoradka*) in the Russian version of the Brockhaus encyclopedia suggested wetland drainage as the most effective means of malaria prevention.[12] "Swamp fever" remained a commonly used term in the Russian Empire, even after scientists had understood the relationship between parasites and mosquitoes that caused the disease.[13]

Fears of malaria were not without justification. Malaria was an endemic disease in the late Russian Empire and the early Soviet state. At the beginning of the twentieth century, it affected some three million people. Infections surged massively during the revolutionary period, when soldiers and refugees helped spread the disease, and during the early 1930s, a time of immense geographical mobility as millions were uprooted by Stalin's attempt to transform the economy and extract resources from rural hinterlands.[14] When scientists had begun to study malaria in the late imperial period, they associated the disease primarily with the empire's southern peripheries, an approach which echoed trends in the international scientific community. Modern malaria science was

[10] Ferdinand Aleksandr Graf fon Garsh, "Otvet na zadachu Imperatorskogo Vol'nogo Ėkonomicheskogo Obshchestva," *Trudy Vol'nogo ėkonomicheskogo obshchestva* ch. 51 (1796): 58.

[11] For instance, "Russkie bolota," *Vestnik Evropy* 8, no. 12 (1873): 754–773, 759.

[12] "Bolotnaia likhoradka (maliariia)," in *Ėntsiklopedicheskii slovar'*, Vol. IV (Saint Petersburg: Brokgauz and Efron, 1891), 311.

[13] N. K. Kol'tsov, *Bolotnaia likhoradka (Maliariia) i komary* (Moscow: Tipogr. Lomonosov, 1913).

[14] Mary Schaeffer Conroy, "Malaria in Late Tsarist Russia," *Bulletin of the History of Medicine* 56, no. 1 (1982): 41–55; Leonard J. Bruce-Chwatt, "Malaria Research and Eradication in the USSR: A Review of Soviet Achievements in the Field of Malariology," *Bulletin of the World Health Organization* 21, no. 6 (1959): 737–772.

entrenched in Europe's colonial history and carried traits of a broader discourse about the South as a backward and uncivilized part of the world. In the Russian Empire, these notions were attributed to the Caucasus, Central Asia, and the Black Sea region. Applying hierarchies of colonial thinking, medical experts lamented the alleged cultural backwardness of the non-Russian population in these parts of the empire, seeing it as a factor which, alongside the marshy environments themselves, contributed to high malaria infection rates.[15]

Although malaria was most widespread in the more southern parts of Eurasia, the spatial framing of malaria as a disease of the peripheries was only partly correct. Temperate regions of the empire were affected as well. In the nineteenth century, malaria outbreaks were documented, for instance, in the province of Viatka, northeast of Moscow, in Volhynia, and in parts of Siberia.[16] This explains why fevers featured so prominently in the writings of improvers who thought about the wetlands in central and northwestern Russia and the western borderlands as environments that undermined progress and development. Take the 1879 report by the Ministry of State Domains, which, upon listing various benefits of wetland drainage in the central Russian province of Ryazan, suggested that "the drainage of these swamps will positively affect the health of the people in the region. In many places, the fevers never stop, which naturally comes from the harmful vapours emitted by the swamps."[17] While late imperial experts, much like their counterparts abroad, framed malaria as a "tropical disease," the actual geography of malaria defied their attempts to cast it as a problem rooted in the fringes of the empire.

With the rise of peat extraction in the late imperial period, peat workers were regularly exposed to malaria. A 1903 Zemstvo survey on malaria in the province of Moscow noted that the disease was most widespread in areas with a large population of factory workers and near peat extraction sites or brickworks. In the Bogorodsk district, where several industrial enterprises had extracted peat since the late nineteenth century, malaria rates were particularly high. In the Vasilevskaia volost, peat workers were the most affected professional group, accounting for a

15 Matthias Braun, "Von Menschen und Mikroben: Malaria im Zarenreich und der Sowjetunion in globaler Perspektive," in Martin Aust, ed., *Globalisierung imperial und sozialistisch: Russland und die Sowjetunion in der Globalgeschichte 1851–1991* (Frankfurt a. M.: Campus, 2013), 353–372, 358–363.

16 Schaeffer Conroy, "Malaria," 42.

17 "Osushenie bolot," in *Sbornik svedenii po departamentu zemledeliia i sel'skoi promyshlennosti* (Saint Petersburg: Ministerstvo Gosudarstvennykh Imushchestv, 1879), 119.

quarter of the documented 1,746 malaria cases in 1898 and 1899.[18] Health provisions for peat workers were generally treated as a secondary issue, and their quality remained poor. However, in those cases where companies or landowners overseeing peat extraction works put such services in place, malaria was a commonly reported disease, while quinine, then a widespread malaria medication, ranked among the most frequently prescribed medicines.[19]

The problem gained traction after the Bolshevik takeover. At the peak of the early Soviet malaria wave in 1922 and 1923, when the authorities registered more than five million infections, central Russia's industrially exploited peatlands became veritable malaria hotbeds. In 1922, the disease affected more than half of all peat workers in the region of Moscow. Some extraction sites noted infection rates of up to 80 or even 90 percent. In the following year, 31.2 percent of the close to 42,000 peat workers in the region of Moscow contracted malaria. Given that public health institutions at that time struggled to generate coherent statistics and mild cases could easily remain undetected, this number possibly underestimated the actual rate of infections. The peat works of Ėlektroperedacha and Shatura, which featured prominently in the early Soviet campaign to mobilize local fuels for the nascent power industry, ranked among central Russia's most severely affected areas. The public health situation was so urgent that it caught the attention of central institutions. State officials feared the peat industry would fall behind production targets with so many sick workers. In 1923, the People's Commissariats of Health and Agriculture and the VSNKh decreed that peat companies should invest a certain amount of money in preventing and treating malaria for each *pud* of peat they extracted.[20]

The issue, however, was not just that peat workers suffered infections in the context of a larger malaria epidemic. In fact, peat extraction was one of the factors that contributed significantly to the surge in infections in central Russia. In 1947, medical expert L. I. Oganov, who had been

[18] *Sbornik statisticheskikh svedenii po Moskovskoi gubernii. Vypusk II: Raspronstranenie i proiavlenie maliiarii v naselenii Moskovskoi gubernii*, ed. A. I. Skibnevskii (Moscow: Pechatnia S. P. Iakovleva, 1903), 16–17, 21, 67.

[19] "Meditsinskaia pomoshch' na torfianykh razrabotkakh," *Vestnik Torfianogo Dela* no. 2 (1916): 302–303.

[20] I. Dobreitser, "Bor'ba s maliariei na torforazrabotkakh," *Izvestiia Narodnogo Komissariata Zdravokhraneniia* 6, no. 4–5 (1923), 13–14; V. I. Martsinovskii, *Maliariia v torfodobyvaiushchei promyshlennosti i bor'ba s nei* (Moscow: TsUTorf, 1923), 9, 12, 21–22.

involved in malaria prevention since the 1920s, submitted a doctoral thesis on malaria in the region of Moscow. Reviewing the existing scholarship, Oganov stated that the causal relationship between peat extraction and high malaria rates was "so firmly established that it does not require proof anymore."[21] The reason lay in the unique constellation between humans, mosquitoes, and water at peat extraction sites. As French and Italian scientists discovered in the late nineteenth century, malaria is caused by a parasite and transmitted through *Anopheles* mosquitoes, which act as disease vectors carrying the parasite from infected to healthy hosts, whether humans or animals. Historically, malaria epidemics often coincided with high levels of human mobility and landscape changes which created breeding grounds for *Anopheles* near large host populations without immunity.[22]

Central Russian peat extraction sites fit this pattern for two reasons. First, peat workers were highly mobile. The seasonal nature of extraction meant that thousands of peasants traveled between their home villages and the peat works every year. Staying in crowded accommodations in the immediate environs of the peatlands, these workers were an easy target for *Anopheles* mosquitoes, which feasted on the large number of human bodies arriving in the spring. The landscape transformations caused by the peat industry were the second major factor contributing to the unprecedented spread of malaria in central Russia. Before the milled peat extraction method proliferated, mechanized extraction using excavators and the hydro-peat method created numerous shallow quarries. Once excavation paused or ended in a particular spot, these quarries filled with water and were colonized by water-loving plants such as mosses or sedges, which made them excellent breeding sites for malaria-bearing mosquitoes (Figure 5.1).[23]

[21] L. I. Oganov, "Maliariia v Moskovskoi oblasti i ee sezonnaia periodichnost'" (Dissertatsiia na soiskanie uchenoi stepeni kandidata meditsinskikh nauk, Moskovskii oblastnoi imeni Mechnikova institut ėpidemiologii, mikrobiologii i infektsionnykh boleznei, 1947), 19.

[22] Timothy Mitchell, *Rule of Experts: Egypt, Techno-Politics, Modernity* (Berkeley: University of California Press, 2002), chapter 1; John R. McNeill, *Mosquito Empires: Ecology and War in the Greater Caribbean, 1620–1914* (Cambridge: Cambridge University Press, 2010), 52–57, 198–234; Frank M. Snowden, "Mosquitoes, Quinine and the Socialism of Italian Women 1900–1914," *Past & Present* no. 178 (2003): 176–209.

[23] E. I. Martsinovskii and V. I. Martsinovskii, "Torfodobyvaiushchaia promyshlennost' i maliariia," *Gigiena i Ėpidemiologiia* 2, no. 1 (1923): 60–67.

FIGURE 5.1 Hydropeat extraction sites near Shatura, 1928. *Source:* Russian State Film and Photo Archive. Reproduced with permission.

During the 1920s, the peatlands of central Russia turned into laboratories of early Soviet malariology. In part, it was here that scientists and medical workers studied the nature of the disease and tested place-based and socio-economic approaches to control it. Soviet health experts and representatives from Eastern European countries advocated these approaches in the League of Nations, often combining emancipatory approaches to public health with paternalistic takes on minorities or subaltern population groups.[24] On the ground, Soviet medics developed a syncretic strategy targeting both the vector of the disease and its human hosts. Following the surge in malaria cases in the early 1920s, malaria stations or seasonally deployed health teams began to enact local campaigns against the disease. Serving two peat-fueled power plants east of Moscow, the peat works in Shatura and Ėlektroperedacha received

[24] Bogdan C. Iacob, "Malariology and Decolonization: Eastern European Experts from the League of Nations to the World Health Organization," *Journal of Global History* 17, no. 2 (2022): 233–253; Braun, "Von Menschen und Mikroben," 363–366.

permanent malaria stations, which coordinated activities to alleviate the situation (Figure 5.2). Their staff carried out general medical examinations and tested the blood of arriving workers to identify those carrying the parasite in their bodies, prescribed and controlled the use of quinine by infected workers, and hospitalized patients with severe symptoms. Public health workers also distributed mosquito meshes and tried eliminating open water containers or puddles. In addition, malaria stations used sulfur fumigation (*okurivanie*) of barracks or sprayed kerosene onto the surface of standing waters (*neftevanie*) to exterminate or dispel the mosquitoes. Complementing these efforts were educational events teaching workers the dangers of malaria and methods to prevent its spreading.[25]

But focusing the campaign on peat extraction sites was not enough. The disease spread in multiple directions, circulating between peatlands and the villages from where the seasonal workers came and to which they returned at the end of the season. Most infections occurred during the summer when extraction was in full swing. However, when Soviet health experts began to study the malaria wave in central Russia, they realized that quite a few peat workers arrived at the extraction sites with an infection. During the early 1920s, more than 90 per cent of seasonal peat workers in the region of Moscow came from the neighboring Ryazan, where, in some areas, peasants relied quite heavily on hand-cut peat for domestic use. Motivated by the lack of firewood, hand-cutting created conditions highly conducive to malaria transmission. Particularly in villages, where peasants extracted peat over multiple years, they left numerous small pits filled with water and plants, which attracted *Anopheles* mosquitoes. When infected peasants from such villages traveled to peat extraction sites and were bitten by an *Anopheles* mosquito breeding in an adjacent drainage canal or peat quarry, the disease would spread as the malaria-bearing mosquito attacked other people. Peat workers who had arrived healthy then carried the parasites to their home villages, where the disease would subsequently take root. Aware of this cycle, central institutions deployed money, equipment, and staff to implement anti-malaria measures in villages known for sending large numbers of people to extract peat every year. During winter, malaria stations or teams tested the local

[25] V. I. Martsinovskii, "Itogi chetyrekhgodichnoi bor'by s maliariei na torforazrabotkakh," in *Trudy 1-ogo s-ezda vrachei torforazrabotok 8–10 Ianvaria 1926* (Skopin: Narkomzdrav-Gostorf, 1926), 6–27; "Protokol zasedaniia komissii po voprosu o vyrabotke mer po sanitarnoi okhrane na otrabotannykh torforazrabotkakh" (1924), RGAĖ f. 478, op. 7, d. 2671, l. 52, 54.

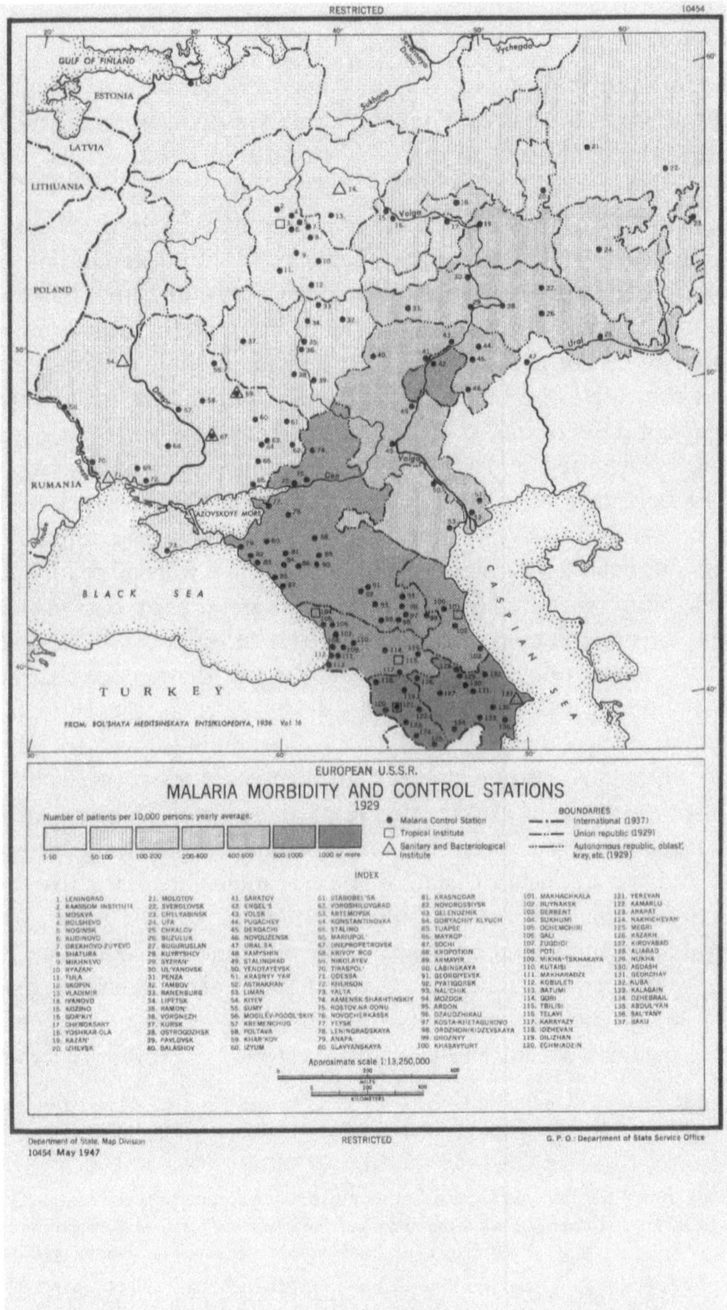

DECLASSIFIED Authority E.O. 13526
NND 76630

FIGURE 5.2 Malaria morbidity and control stations, European U.S.S.R., 1929. Created by the US Department of State in 1947, the map shows a cluster of malaria stations to the east and southeast of Moscow where malaria infections abounded at or near peat extraction sites. Low morbidity rates in central and northwestern Russia indicate low rates of tropical malaria. *Source:* Courtesy U.S. National Archives (NAID 159081167).

population and treated infected peasants with quinine. Although they could not reach all peasants employed as seasonal workers, they still identified and cured several thousand malaria infections in the Ryazan region during the late 1920s.[26]

Malaria did not strike everywhere and everybody in the same way. Early Soviet medical experts developed a nuanced understanding of the geography and demography of the epidemic, revealing how human and nonhuman factors interacted to shape its course. To their relief, most infections in central Russia were of the relatively mild *malaria tertiana* type. In contrast, the much more severe *malaria tropica*, which raged in the southern parts of the Soviet Union, accounted for less than 3 percent of all cases reported in 1925.[27] However, when peat companies hired workers from regions in which *malaria tropica* was more common, its share in overall infections tended to increase, reinforcing colonial anxieties about malaria as a disease originating at and transmitted by people from the Southern peripheries.[28] Contemporary experts also saw a link between the malaria epidemic and the treatment of peat workers, identifying hard labor, poor food, and overcrowded accommodation, where healthy and infected workers slept side by side, as the chief factors facilitating transmission. According to their findings, women showed consistently higher infection rates than their male co-workers and were also disproportionally affected by tropical malaria. This was a direct result of the gendered labor division, as women mostly took care of the drying and the stocking of peat. While the immediate extractive works, dominated by male workers, lasted only until July or August, women's responsibilities often continued into September and October. As a result, women were exposed to the risk of contracting malaria much longer than men.[29] Soviet experts could have diverging views about how best to combine medical, social, and educational efforts to prevent or eradicate

[26] L. I. Oganov, "Bor'ba s maliariei na meste postoiannogo zhitel'stva torforabochikh," in *Trudy vtorogo soveshchaniia po voprosam bor'by s maliariei sredi torforabochikh, 11–13 fevralia 1928* (Moscow: Gosudarstvennoe meditsinskoe izdatel'stvo, 1928), 50–55.

[27] Martsinovskii, "Itogi," 20.

[28] V. K. Sumilov, "Dvizhenie tropicheskoi maliarii i mery bor'by s nei," *Trudy 1-ogo s-ezda*, 48–52.

[29] S. Aigin, "Zarazhaemost' maliariei na torforazrabotkakh i ee prichiny," *Trudy 1-ogo s-ezda*, 27–35; "Perepiska po povodu torfianoi promyshlennosti" (1926–27), RGAĖ, f. 4372, op. 5, d. 377, ll. 19–20; Sulimov, "Dvizhenie tropicheskoi maliarii," 51.

malaria.[30] The picture emerging from their reports, however, is clear. Gender and social inequality embedded in the setup of the peat industry contributed to and reinforced the vulnerabilities linked to the conversion of peatlands into industrial extraction sites.

Keeping or making workers healthy was a function of Soviet productivism, not a goal in itself. Although health experts understood how social and ecological conditions reinforced each other to help spread malaria, there was only so much they could or were willing to do. Contemporary observers tended to see malaria first and foremost as an economic problem. In 1923, a *Pravda*-article declared malaria the country's "new enemy," listing the loss of working hours in the peat industry among the many reasons making the disease so dangerous for the Soviet project.[31] Health experts were not immune to this reasoning and combined medical statistics with information about the economic implications of a wider epidemic, such as health insurance payments, medical bills, and an overall reduction in productivity.[32] Aligning with the government's commitment to mobilize local fuels for the industrializing economy, they worked toward maintaining the peat industry in an operational state. However, neither on the ground nor in central institutions did medical professionals possess enough leverage to force peat companies to improve workers' labor and living conditions. Peat companies, on their part, had little incentive to do so, given that peasants continued to appreciate their seasonal employment in peat extraction as a way to increase their household income. Moreover, despite the state's growing efforts to oversee and control workers' mobilities, peat workers' recruitment and travel routes remained unplanned. Malaria rates decreased notably after peaking in the early 1920s but the disease stayed endemic in peat-extracting regions.[33]

Malaria declined sharply in the Soviet Union in the decade following the Second World War.[34] In central Russia, several factors coalesced to break the fateful link between extraction, mobility, and disease, which

30 Kikimorin, "Otsenka otdel'nykh protivomaliariinykh meropriiatii v usloviiakh raboty na torfu," *Trudy 1-ogo s-ezda*, 54–9; "Preniia po dokladu Kikimorina," ibid., 59–63; Dodonova, "Metody lecheniia maliarii v usloviiakh raboty na torfe," ibid., 73–78.

31 M. Semashko, "Maliariia – novyi vrag," *Pravda* (August 2, 1923): 1.

32 Martsinovskii, "Itogi," 23–25.

33 Oganov, "Maliariia," 66, 79–80; "Spravka o vypolnenii Gosudarstvennogo Plana po torfianomu fondu Vologodskoi oblasti na 1948 god po Upravleniiu torfianogo fonda pri Oblispolkome," GARF f. A-412, op. 2, d. 31, ll. 150–160.

34 Bruce-Chwatt, "Malaria Research and Eradication," 740.

had marked peat extraction sites for many years. With the gradual adoption of the milled peat extraction method, which required much more profound land drainage, peat extraction sites turned from mosquito habitats to arid landscapes, which had very little in common with the rich and ecologically complex wetland ecosystems they had once been. Milled peat extraction also reduced the peat industry's need for seasonal labor and, thus, the number of potential human malaria hosts. Peat trusts' rising investments in workers' settlements further helped slow malaria transmission. However, the end of the malaria epidemic did not make peat extraction sites less hazardous. As time went on, industrial extraction accentuated another characteristic of the specific material environments in which it took place: peat's ability to burn.

CONFRONTING SMOKE AND FIRE

Fires often exposed the hazards created by human interference in peatland ecosystems. Peatlands can burn without humans being involved, but they are much more likely to catch fire when drained. Unlike wood, peat tends to smolder rather than burn with an open flame, which produces dense and smelly smoke. Peat fires often go unnoticed and spread through the soil, a process which can continue even when the ground is covered by snow. The smoke of burning peat played an important role in people's daily interactions with peatland environments and in the broader cultural imagination. In Vladimir Nabokov's first novel Mary (*Mashen'ka*), the smoke of burning peatlands accompanies a young man as his train carries him away from his former lover and into an unknown future: "the train clattered on between peat-bogs burning in the tawny torrent of the sunset; the greyish peat smoke drifted gently over the ground, forming what seemed like two waves of mist between which the train clove its way."[35] For those who experienced it, the smoke from peat fires bore little resemblance to this melancholic moment in Nabokov's novel. During large fire events, it could travel hundreds of kilometers. In the hot and dry summer of 1972, when massive wildfires raged through central Russia, Gorbachev's future advisor on foreign affairs, Anatoly Chernyaev, noted in his diary: "Again day after day 35–36°. And the peat is burning somewhere near Shatura and all of Moscow (and the Moscow

[35] Vladimir Nabokov, *Mary*. Translated from the Russian by Michael Glenny in collaboration with the author (New York: Vintage International, 1989), 78.

region) is covered in a blue veil of smoke. The sun does not even break through it."[36]

That drainage and extraction make peatlands more vulnerable to fire was long known. In 1852, an article published in the *Transactions* of the Free Economic Society promoted the use of peat fuel among peasants. Explaining how to dig drainage ditches and build ovens suited to burn peat, the author warned that peatlands, especially once drained, had to be carefully protected from fire. That peat fires presented a particular challenge because burning peat absorbs water only poorly, was known at the time as well: "If during a dry summer, out of imprudence, a peatland starts burning, it is very difficult to stop the fire which can burn several months, even during the winter under the snow."[37] The willingness to accept such risk reflected the economic gain expected from the drainage and exploitation of peatlands, as well as a broader elitist discourse that marked fire as an indicator of backwardness. In rural contexts, where wooden buildings could burn down rapidly, leaving families and, in severe cases, entire villages without homes, fires seemed to confirm a widespread stereotype about peasants' inadequate way of life and their rootedness in a culture of ignorance and irrationality.[38] Since the ability to control fire appeared as a marker of progressiveness in this broader debate, it is not surprising that advocates of wetland drainage, whose entire mission revolved around the ambition to make nature controllable, took a cavalier attitude toward fire. An official publication from 1896, which reviewed state-led reclamation in central Russia, dismissed fears that drainage would make the land more susceptible to fire as an expression of "overly narrow views." Like resistance to modern technologies such as the railways or the telegraph, drainage criticism, the authors suggested, failed to recognize the "achievements of the human genius" and that progress always demanded "some victims and temporary material losses."[39] Fires in this reading were a necessary evil, a small tribute on the path to modernity.

[36] Anatolii S. Cherniaev, Diary, 1972, *Prozhito*, https://corpus.prozhito.org/note/61206. Accessed March 25, 2025.

[37] "Nastavlenie dlia krest'ian o tom, kak dobyvat' torf i kak ustroit' pech' dlia topki syrym torfom, po sposobu pochetnago grazhdanina Zubchaninova," *Trudy Vol'nogo ėkonomicheskogo obshchestva* 1, no. 2 (1852): 34.

[38] Cathy A. Frierson, *All Russia Is Burning: A Cultural History of Fire and Arson in Late Imperial Russia* (Seattle: University of Washington Press, 2002), 4–7, 51–57.

[39] "Zapadnaia ėkspeditsiia po osusheniiu bolot," *Kratkii ocherk rabot ėkspeditsii po osusheniiu bolot v tsentral'nom raione* (Saint Petersburg: Zapadnaia Ėkspeditsiia po osusheniiu bolot, 1896), 28–29.

When a fire broke out, contemporaries quickly noted peatlands' distinct way of burning. In 1910, the Ministry of Agriculture received a report from a turf master describing his experience with peat fires in the Moscow region. The official had many fire events on the record. Once, during the 1890s, thousands of workers were mobilized to fight a fire at the peat works of the Nikol'skaia cotton mill, which had been burning for several weeks. In other cases, peat fires could be contained soon after being detected. Thus, when the state-owned Biserovskoe peatland caught fire in 1910, the local forester managed to prevent a larger disaster by deploying three fire brigades and some hundred peat workers and peasants from surrounding villages. Still, even when brought under control, peat fires posed particular challenges, which clearly distinguished them from forest fires: "Extinguishing a peat fire is complicated by the fact that the peatland does not actually burn, but charring (*obuglivanie*) occurs." Working under the trying conditions of a peat fire for an extended time, firefighters often suffered headaches, vomited, or fainted. Holes in the ground filled with hot ashes presented an additional danger for firefighters, who risked severe burns when stepping into them.[40]

Other than the turf master, who underscored the specific nature of peat fires, the central administration considered them a primarily financial issue. The local official's report was part of a larger conversation involving the Ministry of the Interior and the Ministry of Agriculture about how people called to fight peat fires on state or private land were to be remunerated. Officials in both ministries agreed that the seriousness of peat fires and the benefits of an immediate response justified considerable financial compensation for firefighters.[41] While this was undoubtedly an important decision, the reasoning in the central bureaucracy underscored the widespread notion of fire as a management issue. The fact that the disturbance of peatlands through drainage and extraction drastically increased the risk of fire was not even brought up.

Peat fires also accompanied the construction of electric power stations. Contemporary and historical accounts of central Russia's first regional power plants emphasize the politics and technologies of electrification, but their history also provides clues about how drainage and extraction made tension and irritation crucial elements of human–peatland

[40] Letter from the turfmaster in Reutevo to the Department of Agriculture and Land Improvement (1910), RGIA f. 425, op. 1, d. 35, ll. 34–35.

[41] Department of Agriculture and Land Improvement, Internal correspondence on peat fires (1910), RGIA f. 425, op. 1, d. 35, ll. 31–32.

interactions in central Russia. In the summer of 1912, when the construction of the Ėlektroperedacha power plant had just begun, a fire raged through the peatland, which was meant to supply the future utility, and through the surrounding forests.[42] Particularly dramatic was a fire in Shatura, a flagship of the early Soviet electrification campaign. At the end of the 1920 season, shortly before most workers were due to travel back to their home villages, the flames undid much of what had been accomplished at the site since 1918, when peat extraction and the building works for the preliminary Shatura power station had started. Workers, peasants, fire brigades, and Red Army soldiers tried to get the fire under control. However, as one worker would remember later, "they were not familiar with peat fires," and their attempts to deal with the trying situation seemed rather chaotic.[43]

Once extinguished, the fire received a firm place in the popular Shatura myth. A worker's poem from 1921 commemorated the fight against the fire as an act of heroism that indicated that a bright future was waiting for Shatura:

For eight days did the fire burn,
You couldn't see anything.
In the end the fire stopped,
And the work began again.

Let us forget about the fire
If only the CENTRAL (the Shatura power plant – KB) is intact.
The barracks and the settlements
We will be build forever![44]

Later official accounts underscored how soon after the fire constructions moved ahead. Still, the toll it had taken was considerable. Some 30,000 metric tons of peat, almost half of what had been extracted in 1920, was lost to the flames. So were 43 houses, several newly built workers' barracks, railway sleepers, and machinery.[45] Even though fire events could easily be mobilized for tales of progress and human willpower, they unmistakably exposed the vulnerabilities emerging from the intentional disturbance of peatlands for the power industry.

42 R. Ė. Klasson, Memoirs (early 1920s), RGAĖ f. 9508, op. 1, d. 9, l. 6, Simonov, *Razvitie ėlektroėnergetiki*, 169.

43 "Vospominaniia o Shature," GARF f. 7952, op. 3, d. 459, ll. 45–46.

44 Worker's poem on the 1920 Shatura peat fire (1921), RGAĖ f. 9455, op. 1, d. 278, l. 60.

45 *Shatura: Sbornik statei k 25-letiiu GRĖS imeni V. I. Lenina* (Moscow, 1951), 10; Zaikov, *Geroicheskaia Shatura*, 14–16.

Not surprisingly, fire was a continuous concern wherever peat was extracted. Labor contracts usually contained sections prescribing how peat workers ought to behave during a fire, while planning authorities issued fire safety regulations for the peat industry. Fire safety also became a topic in the specialized literature on peat extraction.[46] Still, even though the problem was known, fires occurred regularly and often with severe consequences. To the frustration of planners and managers, such events interfered with the economic targets peat trusts were supposed to meet. In 1943, for instance, Shaturtorf lost over 12,000 metric tons of peat due to fire.[47] Apart from causing physical damage, fires presented an organizational challenge due to their impact on the workflow in the peat industry. Peat trusts had to ensure they had enough staff to check for fire all week round and, when fires broke out, peat workers were sent to extinguish them, which meant they had to neglect their other tasks.[48] From the authorities' point of view, fire-related damage or underperformance represented a managerial failure. Even though peat companies mobilized means and workers to implement fire safety regulations, administrative complaints about the inadequacy of these measures were common and featured particularly strongly during fire investigations.[49]

Neglect made for an easy explanation, and bureaucrats were happy to embrace it, but it was only one of several factors. The difficulty of matching supply and demand in the Soviet economy, combined with the material peculiarities of the peat industry, arguably presented a crucial reason why peat-extracting companies were so vulnerable to fire. Throughout the Soviet period, objects of production and consumption frequently wound up where they had no use: the harvest went to waste in the fields, machinery was run below capacity for lack of inputs, and

[46] *Trudy Kommissii po izucheniiu torfianogo dela*, 114; *Artel'nyi dogovor ob usloviiakh oplaty i poriadke rabot torfianikov v sezon 1925 goda* (Moscow: VSNKh – Gostorf, 1925), 5; "O meropriiatiakh po obespecheniiu mer pozharnoi bezopasnosti" (March 1939), RGAĖ f. 7969, op. 1, d. 21, ll. 103–104; "Prikaz No. 29 po Glanomu Upravleniiu torfianoi promyshlennosti Narkomata Ėlektrostantsii SSSR" (1942), RGAĖ f. 7969, op. 2, d. 5, ll. 13–42; G. A. Raschetin, *Profilaktika i tushenie torfianykh pozharov* (Moscow: Stroiizdat, 1964).

[47] Shatura Peat Trust, Annual report (1943), RGAĖ f. 7969, op. 2, d. 241, l. 79.

[48] "Spravka o perevode predpriiatii i organizatsii na 5-ti dnevnuiu rabochuiu nedeliu s dvumia vykhodnymi dniami" (1966), GARF A-632, op. 2, d. 154, ll. 14–19; Report of the Baksheevskoe Peat Company, Shaturtorf (1967), GARF f. A-632, op. 2, d. 158, l. 49.

[49] For instance, "Otchet o rezul'tatakh kontrol'ia za ėksploatatsiei torfianykh mestorozhdenii" (1947), GARF f. A-412, op. 1, d. 61, l. 3; "O neudovletvoritel'nom sostoiianii pozharnoi bezopasnosti i proizvodstvennoi distsipliny na predpriiatiach Sverdlovskogo torfotresta (1969)," GARF f. A-632, op. 2, d. 277, ll. 1–3.

readily produced commodities were stockpiled in storehouses.[50] Logistical problems created multiple knock-on effects, often leaving workers, managers, and consumers frustrated because they could not get on with their duties or buy what they needed. However, they created a severe hazard in the peat industry, where the unused object was highly flammable. Over decades, companies struggled to transport and deliver their peat at the end of the season. Finding buyers could also be a problem. Consequently, extracted peat often remained unused for months, sometimes still being on the fields when the new extraction season began in the spring.[51] With the adoption of the milled peat extraction method, this turned into more than an economic issue (Figure 5.3). Usually stored in piles a few meters in height, milled peat is particularly prone to self-heating ignition (*samovozgoranie*), a phenomenon caused by microbiological and chemical reactions in peat. Smoldering often undetected for days or weeks, such fires regularly destroyed several months' work. From the 1930s, when the peat trusts first began experimenting with milled peat extraction, losses due to self-heating ignition appeared periodically in the economic record.[52]

Heat, drought, and wind further added to the fire hazard, making it almost impossible to control the complex material assemblages in the peat industry. Peat-extracting enterprises were expected to isolate piles of milled peat from each other to prevent fires from spreading, but even when they tried, they could barely keep up with this task. In the spring of 1967, several peat companies of the Shatura, Yaroslavl, and Sverdlovsk Peat Trusts suffered considerable losses because the peat from the previous season went up in smoke.[53] Shaturtorf's company Solotchinskoe experienced dramatic times after several peat piles in the company's

[50] Natalya Chernyshova, *Soviet Consumer Culture in the Brezhnev Era* (London: Routledge, 2013), 22–26; Serguei A. Oushakine, "'Against the Cult of Things': On Soviet Productivism, Storage Economy, and Commodities with No Destination," *The Russian Review* 73, no. 2 (2014): 198–236; Martin Kragh, "The Soviet Enterprise: What Have We Learned from the Archives?," *Enterprise & Society* 14, no. 2 (2013):360–394, 379–380.

[51] See, for instance, "Ob-iasnitel'naia zapiska k godovomu otchetu za 1967 god po osnovnoi deiatel'nosti Shaturskogo gosudarstvennogo tresta," GARF f. A-632, op. 2, d. 192, l. 87.

[52] Dlugotskii et al., *Frezernyj torf*, 41–43; Z. P. Kuzhman, "Klassifikatsiia polei frezernogo torfa po sklonnosti ego k samovozgoraniiu," *Torfianaia promyshlennost'* 30, no. 9 (1953): 5–7; Shatura Peat Trust, Annual report (1939), RGAĖ f. 7969, op. 1, d. 269, l. 63.

[53] "O podgotovke torfopredpriiatii k sezonu dobychi torfa 1968g." GARF f. A-632, op. 2, d. 107, ll. 128–129.

FIGURE 5.3 Milled peat extraction in the region of Moscow (1959). *Source*: Russian State Film and Photo Archive (RGAKFD). Reproduced with permission.

extraction fields had begun to smolder at the end of the 1967 peat season. Strong wind created a massive fire, which soon spread beyond the company's territory and into the surrounding forests. Within days, more than 25 hectares of extraction sites and 50 hectares of forest were burning, while people living in a neighboring village feared for their homes. It took some 2,000 workers and army recruits to bring the fires under control.[54] A similar scenario unfolded in Shaturtorf's company Petrovskoe the following year, as heat and gusty winds coalesced, turning a few initially tiny fires caused by spontaneous ignition into a blaze which soon encroached on a nearby settlement. Fortunately, the fire could be localized. It still led to considerable damage, however, destroying two houses and three barracks.[55]

THE ENDLESS SUMMER OF 1972

Fires such as the ones in the Solotchinskoe or Petrovskoe companies were dramatic. But they paled compared to the inferno of 1972, when a

[54] "O pozhare na Solotchinskom torfobriketnom predpriiatii" (1967), GARF f. A-632, op. 2, d. 107, ll. 137–139.

[55] "O pozharakh, proisshedshikh na riade torfopredpriiatii Shaturskogo, Iaroslavskogo, Ivanovskogo, Vladimirskogo, Orekhovskogo i Leningradskogo trestov" (1968), GARF f. A-632, op. 2, d. 202, ll. 67–71.

prolonged drought and heat wave during the summer created a fire emergency in large parts of central Russia. For weeks, wildfires raged through forests and peatlands that had been turned into extraction sites. Their scale was massive. More than 65,000 hectares of forests and almost 5,000 peat piles burned in the region of Moscow. Seventy thousand people, among them more than 20,000 army recruits, tried to get the situation under control.[56] Given that information about disasters was heavily censored at the time, most Soviet citizens only had a very vague idea of what was going on. Dmitrii Oreshkin, who was a geography student in 1972, remembered decades later that "nobody knew that several villages burnt down, that more than one hundred people died and that Marshal Grechko (then Soviet Minister of Defense – KB) personally led the fire defense from Shatura."[57] Contemporary sources paint a similar picture. On August 8, Soviet dramaturge Alexander Gladkov noted in his diary: "The first report – via BBC – on the burning peatlands in Shatura. Curious how Muscovites learn about such things through the English radio."[58] The following day, the *New York Times* reported that the city of Moscow was cast in thick smoke from burning peatlands in the Shatura region, making people in Moscow suffer from watering eyes and poor visibility. According to the article, the Soviet authorities had remained quiet about the unfolding drama until the news agency TASS, apparently in response to the smoke in Moscow and international media footage, issued a brief note.[59]

Creating an unhealthy environment for millions of people in Moscow, the fires posed a much more immediate and life-threatening danger to humans, plants, and wildlife on the ground. Week after week, the fires disrupted any sense of normality in peat-extracting regions, as they jumped from forests to peat-extraction sites and vice versa. Observers feared for the beautiful nature of the Meshchera Lowland or noted how trees, even when not burning, turned completely dry. Some of

[56] Nikolai Dronin and Edward Bellinger, *Climate Dependence and Food Problems in Russia 1900–1990: The Interaction of Climate and Agricultural Policy and Their Effect on Food Problems* (Budapest: CEU Press, 2005), 245–249; "Lesnye i torfianye pozhary leta 1972 g. Spravka," https://ria.ru/20100805/262042285.html?ysclid=lhzc4gayby525403981. Accessed March 29, 2025.

[57] "Nikto ne znal, chto sgorelo neskol'ko dereven' i pogiblo bol'she sta chelovek," *Kommersant* (July 11, 2014), www.kommersant.ru/doc/2503994?ysclid=lhkpgltpqq328567655. Accessed July 15, 2024.

[58] Aleksandr Gladkov, Diary, August 8, 1972, *Prozhito*, https://corpus.prozhito.org/note/114789. Accessed March 25, 2025.

[59] "Smoke Shrouds Moscow as Peat-Bog Fire Rages," *New York Times* (August 9, 1972), 2.

Shaturtorf's companies had to entirely cease their activities, as all workers were called up to fight the blaze. The trust's newspaper *Za Torf* cheered workers, firefighters, and the wider population as they were trying to protect their homes, workplaces, and the surrounding forests from the flames. In keeping with the conventions of Soviet hero narratives, numerous articles evoked the dedication and the stamina that people demonstrated in their fight against the "insidious elements" (*kovarnaia stikhiia*). But these cheerful narratives could not hide that the situation was serious. Workers got injured during their attempts to extinguish the fires; in Shaturtorf, the flames threatened to encroach on the telephone line, which would have left the settlement without a direct connection to the town of Shatura, where the firefighting operations were coordinated. In another case, women and children were temporarily evacuated from their homes because the flames got so close to their settlement. The workers of the Shatura Peat Trust were luckier than some of their colleagues elsewhere, as no houses were lost during the frightening weeks of August and September. The events, however, took a sad turn when 19-year-old Liubov' Arefeva and 23-year-old Anatolii Kiniazev died from smoke inhalation while fighting the fires on one of Shaturtorf's extraction fields.[60]

Who was to be blamed? This question arose as soon as the magnitude of the fires became apparent. Although the drought of 1972 was known to have been caused by an abnormally stable anticyclone, there was a general sentiment that humans had played their role in driving the disaster. Contemporary observers understood this responsibility in different ways, however. A convenient shortcut was to blame individuals and their careless attitude toward outdoor fire safety. In dry years, a barbecue in the garden of a summer house, a bonfire at a campsite, or the glowing cigarette stub of a fisherman could easily cause massive wildfires. This was particularly so in forests on drained peat soil, where fires would often smolder and spread through the ground before turning into a visible blaze. While officials and media willingly drew a link between outdoor leisure and the fires, some commentators entertained the possibility that irresponsible behavior was only one part of the equation.[61] In the middle

[60] Shatura Peat Trust, Annual report (1972), GARF f. A-632, op. 2, d. 488, ll. 93–105, l. 111. Shaturtorf's newspaper featured several articles titled "Fighting the elements" during that summer. See "V bor'be so stikhiei," *Za Torf* (August 12, 1972): 2; (August 23, 1972): 2; (August 26, 1972): 2; (August 30, 1972): 1; (September 2, 1972): 2.

[61] See, for instance, "Prikazy Glavtorfa za 1972," GARF f. A-632, op. 2, d. 444, l. 77.

of August, *Literaturnaia Gazeta*, a literary magazine known for its critical takes on environmental matters, featured a long report describing the horrors of the peat fires raging in central Russia. The authors did not hide their impatience with people who portrayed the fires as "natural disasters." They posited that much of the damage and suffering could have been avoided had the authorities prioritized fire prevention and monitoring in the first place. After all, meteorologists had forecast the unusual weather conditions earlier in the year. Apart from raising these structural issues, the article posed even bigger questions. Why was there no scientific literature on how to extinguish peat fires? And weren't the warnings justified that humans' unreasonable interference with nature may yield unpleasant results?[62] The authors provided no answers to these questions. Still, the fact that they articulated them marked 1972 as a crucial moment in the re-evaluation of human–wetland relationships that gradually gained momentum among Soviet scientists and intellectuals.

Shockingly for contemporaries, the danger was not over when the fires were finally under control. During the fall and winter months, the hazards linked to milled peat extraction in an economy where ready produce lay idle became apparent. In November, Glavtorf, the central administration of the peat industry, decided to compensate peat trusts for the losses they had endured because of the fires. Shaturtorf, which had been hit worst, was to receive 5,000 rubles to make up for the destruction of machinery and infrastructure and for the peat that was lost.[63] Probably few people expected that more damage was underway. In December, the first reports arrived about numerous cases of spontaneous combustion in a peat company close to the city of Tver.[64] Other places soon reported the same issue. Fires due to self-heating ignition turned into a massive problem during that winter because the milled peat of the 1972 season had been at a temperature much higher than usual when piled up. Its heightened initial temperature made the peat even more prone to subsequent self-heating and ignition. In several peat companies in the Yaroslavl and Ivanovo regions, which had suffered heavily from heat and fires, workers observed how the temperatures of their milled peat gradually increased between October and December. As the companies struggled to find

[62] Vil' Dorofeev, Valerii Povoliaev and Vladimir Travinskii, "Skhvatka s ognem," *Literaturnaia Gazeta* no. 33 (August 16, 1972): 13.

[63] "O likvidatsii posledstvii pozharov" (1972), GARF f. A-632, op. 2, d. 444, l. 118.

[64] "O likvidatsii ochagov samovozgoraniia na torfopredpriiatii Vasil'evskii Mokh Kalininskogo torfianogo tresta" (1972), GARF f. A-632, op. 2, d. 444, ll. 132–133.

buyers for the peat, workers isolated the piles from one another and tried to reduce their temperature by adding cooler peat. Their efforts were often unsuccessful. One company lost as much as 43 percent of the peat extracted in 1972 due to self-heating ignition. In 1973, representatives of the peat industry approached the central peat administration requesting that hundreds of thousands of metric tons of milled peat, which had gone to waste since the end of the previous season, be written off.[65]

Forgetting and remembering took erratic turns in the fiery history of the peat industry and its aftermath. Measures to prevent fires were as old as industrial peat extraction itself. Yet, even though peat companies gradually learned to manage the risk more effectively, they never entirely eradicated the possibility of fire. This was not only due to their lack of control over atmospheric conditions. Failure to prevent fires often had far more mundane reasons, such as the allocation of resources and responsibilities within the broader economy. During the 1970s, financial problems added to the logistical pressures facing the peat industry. In 1976, it became known that quite a few peat trusts, due to budgetary constraints, had substantially lowered the number of regularly employed firefighters. In a risky bet on the infrequent nature of large fire events, Shaturtorf had reduced their firefighting staff by 15, the Vladimir Peat Trust by 17, and the Kirov Peat Trust by 32 employees. Just four years after the fateful summer of 1972, the number of firefighters had dropped below the norm in several Soviet peat trusts.[66] In light of the recent disaster, the trusts' decision seems a striking example of regressive learning rooted in human arrogance. Still, for a while at least, administrators had good reasons to believe that 1972 had been an exception. In the late Soviet period, peat companies continued to experience fires, but most remained localized events.

Massive peat fires would return only in the new millennium when economic decline and climate change coincided, creating a new form of ecological precarity. In a warming world, central Russia's abandoned peat extraction sites are highly likely to be at the center of major wildfire events, particularly given that fire monitoring decreased substantially after the end of the Soviet period. In 2002, peat and forest fires east of Moscow rekindled the memory of 1972. Commenting on the smoke from

65 "Svodnaia vedomost' inventarizatsii i kachestvennoi kharakteristiki torfa za 1973 god," GARF f. A-632, op. 2, d. 496, ll. 39–50, ll. 54–57; 61–65, ll. 72–73, ll. 87–92, ll. 94–96.

66 "O vosstanovlenii chislennosti rabotnikov pozharnoi okhrany v proizvodstvennykh obedineniiakh" (1976), GARF A-632, op. 9, d. 53, ll. 79–90.

burning peatlands near Shatura, the writer Mark Kharitonov noted in his diary, "old-timers, like me, will remember that it was like this exactly 30 years ago, in 1972."[67] The art critic Sergei Popadiuk equally thought that the smoke in 2002 was "just like in 1972."[68] But 2002 was no exception either. Just eight years later, in the hot and dry summer of 2010, fires once again swept through Meshchera. And, like before, 1972 again served as a crucial historical reference point for journalists and eyewitnesses.[69] However, it was in 2010 that interpretations moving beyond individual neglect or bureaucratic failure slowly began to be articulated. The fact that the disturbance of peatlands lay at the core of the problem increasingly received attention. Even the then-president Dmitry Medvedev, despite the Russian government's and Medvedev's denial of the seriousness of the climate crisis, suggested a link between the fires and the wider planetary predicament.[70]

* * *

Running through the history of human–peatland relations in modern Russia was a fateful irony. While peatlands had been depicted as useless and even harmful since the modern period, they only became truly troubling environments when humans actively destroyed or mismanaged them. The stories of fire and disease in the context of industrial peat extraction demonstrate that peatlands were not just a backdrop to history but could actively undermine or deter human attempts to use them. Ideas and matter interacted in much more intricate ways in the history of peat fuel than it may seem at first glance. Prometheanist visions of nature

[67] Mark Kharitonov, Diary, July 31, 2002. *Prozhito*, https://corpus.prozhito.org/note/198628. Accessed March 25, 2025.

[68] Sergei Papadiuk, Diary, 4 August 2002. *Prozhito*, https://corpus.prozhito.org/note/626347. Accessed March 25, 2025.

[69] "Prirodnye pozhary 1972 i 2010 goda," *RIA Novosti*, August 11, 2010, https://ria.ru/20100811/264161975.html; "Situatsiia so smogom v Moskve pokhozha na 1972 god, schitaet veteran GO," *RIA Novosti*, August 5, 2010, https://ria.ru/20100805/262094678.html; "MChS: nyneshniaia pozharoopasnaia situatsiia slozhnee, chem bylo v 1972 godu," *RIA Novosti*, August 6, 2010, https://ria.ru/20100806/262461618.html. All accessed March 29, 2025.

[70] Viktor I. Danilov-Danil'ian, "Prichiny i uroki torfianykh i lesnykh pozharov 2010 goda," *Ėkologiia i zhizn'* no. 10 (2010), https://elementy.ru/nauchno-populyarnaya_biblioteka/431267. Accessed August 1, 2014. "Pokonchit li zhara v Rossii s ee somneniiami po povodu global'ngo potepleniia?," *INOSMI*, August 3, 2010, https://inosmi.ru/20100803/161816152.html). Accessed July 31, 2024. Chapter 6 discusses the role of fires in abandoned peatlands at greater depth.

domination were engrained in hegemonic ideas and practices of power in Russia's empires. And so they were in modern conceptions and the actual treatment of wetlands. However, as Prishvin's 1926 essay on peat demonstrates, the fact that human relationships with the nonhuman world are never unidirectional was very well understood in Russia, too. Acknowledging the agency of matter and the vitality of peatland ecosystems, the writer recognized the tension between human ambitions and their material constraints. Other thinkers at the time similarly emphasized the dialectics of nature. The Soviet writer Andrei Platonov, for instance, combined a strong commitment to Prometheanist thought with a vision for socialism that was sensitive to how nature worked.[71] That such ideas did not affect the damaging and hazardous interventions in wetland environments was not, then, because they were ignored or dismissed. Gleb Krzhizhanovskii, who had propagated the use of peat after the 1917 Revolution, demonstrated quite a profound understanding of peatlands as living environments while acting as the most important promoter of peat-fueled power plants in the early Soviet state.[72] Instead, it was the alliance, typical for many Soviet industries, between the recognition of nature's doings, the unquestioned imperative of economic growth, and a strong belief in technocratic solutions to environmental problems, which explains the continuous and increasingly destructive exploitation of central Russia's wetlands.[73] The fact that peatlands could turn against their human intruders was accepted as a given and sometimes even served to further justify productivist stereotypes and uses of wetlands. Within this mindset, moments of irritation did not provide strong enough an argument to cease extraction.

The troubled relationships between humans and peatlands in the context of industrial peat extraction point to the logic of power and marginalization operating in Russia's fossil economy. The large-scale exploitation of peatlands was driven by a desire for cheap energy articulated by the political and social elites. In the late imperial period, peat extraction served the vision of supporting central Russia's industries. During the Soviet period, it embodied the energy utopianism of the Bolsheviks and their eagerness to combine economic recovery following the devastation inflicted by war, revolution, and civil war with an agenda

71 Erley, *On Russian Soil*, 93–112.

72 Chapter 2 discusses this question in greater detail.

73 On this "dualistic conception of nature" in Soviet history, see Bruno, *Nature of Soviet Power*, 11–15. This argument is also central to Kochetkova, *Green Power of Socialism*.

for progressive economic transformation. Yet, despite the revolutionary concept of electricity as a tool for emancipation, the power industry itself rested upon and perpetuated historically rooted injustices. The peat from the expansive wetlands in central Russia fueled power utilities such as Ėlektroperedacha or Shatura. From there, electric power moved through high-voltaic transmission wires to large urban and industrial centers (most notably the city of Moscow), away from the environments that served as its material basis. Considering that electricity arrived in the Soviet countryside only in the second half of the twentieth century, it is unlikely that most peat workers ever got to enjoy the actual benefits of electric power in their everyday lives. At the same time, however, they were the people most frequently exposed to malaria and peat fires, dangers resulting from the peat industry's violent disruption of wetland ecosystems. Contemporaries tended to see these problems as collateral damage on a rocky path toward progress, sometimes attributing them to managerial failure. They were less likely to admit, however, that these disasters were both an outcome and an expression of how Russia's fossil economy placed the destruction and hazards it wrought on people and nature at its margins.

6

Revaluation

In December 1978, the Central Committee of the Communist Party and the Soviet Council of Ministers declared that peat extraction for energetic needs would end in the near future. The announcement was part of a comprehensive decree designed to enhance environmental policies and regulations in the Soviet Union. Published under the cumbersome title "On additional measures for the strengthening of nature protection and the improved use of natural resources," it called on administrative, planning, and scientific institutions to embed environmental principles in their day-to-day activities. Thirty paragraphs outlined actions which state institutions were expected to implement to prevent pollution, the overuse of resources, and the destruction of ecosystems. The implications for the peat industry would be significant. The decree instructed the Soviet Ministry of Energy and planning authorities at union and republican levels to establish a timeframe for transitioning peat-based thermal power stations in the European part of the Soviet Union to other types of fuel.[1] After decades of fueling power plants, peat would lose its overall small, but in some regions substantial, role as an element of the Soviet energy system.

To an extent, this decree merely described changes that were already underway. Energy generation, transportation, and consumption in the Soviet Union underwent massive changes in the decades following the

[1] TsK KPSS & Soviet Ministrov SSSR, "Postanovlenie ot 1 dekabria 1978 goda No 984 O dopolnitel'nykh merakh po usileniiu okhrany prirody i uluchsheniiu ispol'zovaniia prirodnykh resursov." Available at https://docs.cntd.ru/document/765700043?marker=1EVOJJH&section=text. Accessed March 27, 2025.

Second World War. The principle of "regional autarky," which the 1921 GOĖLRO-plan had anchored in the country's energy system, gradually gave way to a new spatial setup in which energy traveled across increasingly longer distances. Two factors in particular contributed to this development: the expansion of hydroelectric power generation following the construction of large dams along the Volga, Dnipro, Angara, and Yenisey rivers, and the discovery and exploitation of large oil and gas deposits first in the European part of the Soviet Union and later in Siberia. Nuclear power stations also began producing electricity in the 1970s.[2] As the country's energy system embarked on a trajectory of expansion and interregional integration made possible by a growing network of transmission lines and pipelines, peat's importance began to decline. Peatlands had provided 7 percent of the fuel consumed by Soviet thermal power utilities in 1960, but this share decreased steadily over the following two decades, falling to less than 2 percent in 1980. Unlike coal, whose contribution to fuel use in the power industry decreased in relative but increased in absolute terms, peat use declined in absolute terms. The trajectory was less linear in the household sector, where peat briquette consumption nearly tripled between 1965 and 1975. Peat briquettes went out of fashion once gas became more widely available beyond the cities but kept being used in some rural areas through the Soviet breakup.[3]

While its significance as a fuel was shrinking, peat was redefined in ecological terms. The rise of the environment as a framework for science, politics, and activism in the decades following the Second World War did not bypass the Soviet Union.[4] In the post-Stalinist period, Soviet leaders readily deployed environmental rhetoric, while scientists and writers cautiously articulated ecological concerns in public.[5] As part of this trend,

[2] Gestwa, *Die Stalinschen Großbauten*, 77–113; Frey, *Arktischer Heizraum*, 232–244.

[3] Wilson, *Demand for Energy*, 30–40, 63.

[4] Paul Warde, Libby Robin, and Sverker Sörlin, *The Environment: A History of the Idea* (Baltimore: Johns Hopkins University Press, 2018).

[5] The scholarship on the topic has expanded considerably in recent years. See Melanie Arndt and Laurent Coumel, "A Green End to the Red Empire? Ecological Mobilizations in the Soviet Union and Its Successor States, 1950–2000: A Decentralized Approach," *Ab Imperio*, no. 1 (2019): 105–124; Nicholas B. Breyfogle, "At the Watershed: 1958 and the Beginnings of Lake Baikal Environmentalism," *Slavonic and East European Review* 93, no. 1 (2015): 147–180; Katja Doose, "A Global Problem in a Divided World: Climate Change Research during the Late Cold War, 1972–1991," *Cold War History* 21, no. 4 (2021): 469–489; Laurent Coumel, "A Failed Environmental Turn? Khrushchev's Thaw and Nature Protection in Soviet Russia," *The Soviet and Post-Soviet Review* 40 (2013): 167–189; Marc Elie, "Formulating the Global Environment: Soviet Soil Scientists and the International Desertification Discussion, 1968–91," *Slavonic and East European Review*

the role and the meanings of peat and peatlands in Russia's fossil economy began to change. The fact that the Soviet leadership presented the phasing out of peat as part of a broader array of environmental policies and not as an adjustment to ongoing transformations in the energy sector clearly indicates this. Informed by an ecological understanding of peatlands as complex ecosystems with intimate connections to their surroundings, the 1978 decree stressed that peat resources were not unlimited and should be used responsibly. It also mentioned peatlands' crucial role as regulators of water flows.[6] Carrying key features of late-Soviet technocratic environmentalism, which combined a commitment to economic growth and an awareness of the ecological damage it could inflict with a strong belief in scientific planning, the decree marked a departure from the century-old wetlands discourse.[7] The government admitted that peat would continue to feature as an economic resource but insisted its role would be minimal and solely limited to the farming sector, whose poor performance caused continuous headaches for the political leadership. Still, the ways in which contemporaries framed the value of peatlands had changed quite remarkably. Peat, so long presented as an almost unlimited resource, was now deemed scarce. At the same time, peatlands lost the stigma of being unproductive "wastelands." Instead, they were considered an important part of nature whose use should align with the principles of what late Soviet experts called the "complex" or "rational utilization of natural resources."

It would be misleading to consider the revaluation of peatlands' role in Russia's fossil economy as a straightforward shift from productivism to environmentalism, from exploitation to stewardship, from destruction to recovery. This chapter demonstrates that the process was contested and

93, no. 1 (2015): 181–204; Jonathan Oldfield and Denis J. B. Shaw, "V. I. Vernadskii and the Development of Biogeochemical Understandings of the Biosphere, c.1880s–1968," *British Journal for the History of Science* 46, no. 2 (2013): 287–310; Julia Obertreis, "Von der Naturbeherrschung zum Ökozid? Aktuelle Fragen einer Umweltzeitgeschichte Ost- und Ostmitteleuropas," *Zeithistorische Forschungen* 9, no. 1 (2012): 115–122; Jonathan Oldfield, "Imagining Climates Past, Present and Future: Soviet Contributions to the Science of Anthropogenic Climate Change, 1953–1991," *Journal of Historical Geography* 60 (2018): 41–51; Eglė Rindzevičiūtė, *The Power of Systems: How Policy Sciences Opened Up the Cold War World* (Ithaca, NY: Cornell University Press, 2017).

[6] TsK KPSS & Soviet Ministrov SSSR, "Postanovlenie," paragraph 14.

[7] Laurent Coumel, "Building a Soviet Eco-Power while Looking at the Capitalist World: The Rise of Technocratic Environmentalism in Russian Water Controversies, 1957–1989," in Astrid Mignon Kirchhoff and J. R. McNeill, eds., *Nature and the Iron Curtain: Environmental Policy and Social Movements in Communist and Capitalist Countries 1945–1990* (Pittsburgh: University of Pittsburgh Press, 2019), 17–35.

that the decline of peat extraction led to new forms of social and ecological precarity. The peat fuel industry shrank drastically in the late Soviet period. From over 40 million metric tons in the early 1970s, extraction in the RSFSR declined to 25.9 million tons in 1979 and reached a mere 5.2 million tons in 1990.[8] In the 1980s, the Shatura Power Plant was reconstructed to allow the station to gradually switch to natural gas, while peat extraction in the Meshchera plummeted. After the Soviet collapse, the peat industry lost its vital role as a provider of employment and social welfare in the eastern parts of the Moscow region.[9] Many workers' settlements faced marginalization and outmigration as a result, while the abandonment of peat extraction sites, combined with the state's retreat from the countryside, left many formerly exploited peatlands extremely vulnerable to wildfire, a risk that has been continuously increasing as climate change leads to warmer and dryer summers in central Russia. The role of peat as an industrial fuel is largely forgotten but extraction has an uncanny afterlife in the communities and environments it created.

LIMITS TO GROWTH

Micro and macro, contemporary and present-day perspectives generate different narratives of peat's changing role in Russia's fossil economy during the late Soviet period. With hindsight, the decline of peat appears like an inevitable development: Once the Soviet Union began to tap its gas and oil riches to satisfy growing domestic energy needs and established itself as an energy power internationally, peat, whose energy density was considerably lower, gradually lost its relevance until it disappeared from the energy system almost entirely.[10] The decline of peat fuel did not follow any objective logic, though. Inefficiency, after all, was deeply written into modern energy history.[11] Neither would such a straightforward declensionist narrative match the experiences of contemporaries. In fact, things looked a lot different from their point of view. Following the war, the Soviet economy experienced high rates of recovery growth

[8] *Narodnoe Khoziaistvo v RSFSR v 1972 godu*, 98; *Narodnoe Khoziaistvo v RSFSR v 1979 godu*, 67; *Narodnoe Khoziaistvo v RSFSR v 1990 godu*, 373.

[9] "Shaturskaia GRĖS," www.mosenergo-museum.ru/History_of_Mosenergo/Historical_Review/19077/. Accessed March 27, 2025.

[10] On the rise of the Soviet Union as an energy power, see Högselius, *Red Gas*; Perović, *Rohstoffmacht Russland*; Perović, *Cold War Energy*.

[11] Bonneuil and Fressoz, *Shock of the Anthropocene*, 105–107.

accompanied by increasing domestic fuel consumption.[12] Even when economic growth slowed during the 1960s, managers and peat workers had little reason to believe that peat fuel would soon become a thing of the past. The milled-peat extraction method, which proliferated in many enterprises in those years, facilitated the most significant output increases the peat industry had ever seen. B. N. Sokolov, the lead engineer of the Shatura Peat Trust, boasted in 1971 that the trust would increase extraction by 71 percent during the current Five-Year plan.[13]

Paradoxically, it may seem, peat's relative decline in the energy system at first coincided with a period of notable growth. Sure, the technological advances in the peat industry paled compared to what was happening in oil and particularly in gas extraction. Gosplan's experts had already noted in 1960 that producing a metric ton of "reference fuel" (*uslovnoe toplivo*), a calculation unit used to make different types of fuel comparable, was almost four times as expensive for peat than for natural gas. Even though extraction costs were expected to decrease, peat's relative costliness was projected to rise further because the oil and gas industries would realize significant economies of scale in the coming years.[14] Still, when the planners calculated in 1961 that the share of peat in the Soviet fuel balance would decline to 1.1 percent within the next two decades, they did not envision the end of the peat industry. Given that it was significantly cheaper than coal, peat was thought to retain its competitiveness when used in places close to where it was extracted.[15] In line with this thinking, Gosplan supported several requests to expand extraction in the central Russian regions of Tver, Kirov, Vologda, and Vladimir in 1964.[16] At about the same time, building works started to extend the Shatura Power Plant. When its new unit was completed in 1972, Shatura

12 Mark Harrison, "The Soviet Union after 1945: Economic Recovery and Political Repression," *Past and Present* 210, Supplement 6 (2011): 103–120; G. I. Khanin, "The 1950s: The Triumph of the Soviet Economy," *Europe-Asia Studies* 55, no. 8 (2003): 1187–1211. Khanin uses fuel consumption as one of the proxies to calculate Soviet economic growth in that period.

13 B. N. Sokolov, "Opyt raboty kompleksno-mekhanizirovannykh predpriiatii Shaturskogo torfotresta," *Torfianaia promyshlennost'* no. 3 (1972): 9–11.

14 "K postroeniiu perspektivnogo plana toplivno-ėnergeticheskogo balansa SSSR na 1960–1975gg.," RGAĖ f. 399, op. 1, d. 899, l. 9, ll. 37–38.

15 "Strukturnye sdvigi v proizvodstve i potreblenii topliva (1961)," RGAĖ f. 99, op. 1, d. 56, ll. 1–6.

16 "Porucheniia Soveta Ministrov RSFSR po razvitiiu torfianoi promyshlennosti i dokumenty po ikh vypolneniiu (1964)," GARF A-626, op. 8, d. 3862, ll. 3–4, 11–14, 17–20, 68, 71, 81.

was the largest peat-fueled power plant in the world. At the same time, construction was underway for several smaller peat-based power utilities in the regions of Cherepovetsk, Pskov, and Smolensk.[17]

In some places, however, the tide was already beginning to turn. In 1968, the Soviet Council of Ministers decided to transfer Shaturtorf's peat company Pustoshinskoe from the Ministry of the Fuel Industry to the Ministry of Chemical and Petroleum Engineering, which sought to repurpose Pustoshinskoe's workshops and make them a branch of its expanding Moscow-based pump factory. This small decision was emblematic of the larger structural shifts within the late Soviet economy. The idea of changing the company's affiliation was driven by the growing economic role of the petroleum industry and the government's support for the domestic production of tools and equipment required to extract, transport, and process oil.[18] But Pustoshinskoe's transfer also indicated the mounting economic pressures facing the peat industry in the postwar decades. The company had extracted peat since 1945, though its contribution to Shaturtorf's overall output never reached a significant level. Between 1948 and 1968, the number of employees had fallen from 982 to 225, forcing locals to travel to Ryazan, Roshal, and Shatura for work.[19] When its transfer was decided, Pustoshinskoe had already lost almost every connection to the peat fuel industry. As a short report in the file on the case explained, the company's workers "carried out random works" using raw materials from afar. They made metal closets for locker rooms and mechanical clothes brushes and repaired metal containers for textile companies. Local administrators were probably relieved when the Council of Ministers decided to hand the company and most of its assets to the petrochemical industry. Not only did the central administration lower Shaturtorf's production target for the running year, the redeployment of Pustoshinskoe's workshops also promised to maintain and even increase

[17] Iain F. Elliot, *The Soviet Energy Balance: Natural Gas, Other Fossil Fuels, and Alternative Power Sources* (New York: Praeger Publishers, 1974), 194; N. V. Ivashechkin, "Torfianaia promyshlennost' RSFSR," *Torfianaia promyshlennost'* no. 12 (1972): 2–8.

[18] Viacheslav Nekrasov, "Decision-Making in the Soviet Energy Sector in Post-Stalinist Times: The Failure of Khrushchev's Economic Modernization Strategy," in Jeronim Perovič, ed., *Cold War Energy: A Transnational History of Soviet Oil and Gas* (Cham: Palgrave Macmillan, 2017), 165–199, 191.

[19] "Kvartal'nye otchety torfotrestov po proizvodstvennomu travmatizmu po torfopredpriiatiiam za 1948 god," RGAĖ f. 7969, op. 2, d. 640, l. 73.

local employment, thus avoiding the social cleavages that a gradual winding down of the enterprise would have entailed.[20]

But what exactly was the matter? That Pustoshinskoe served other industries instead of focusing on peat-based products and maintaining Shaturtorf's machinery was the result of a rocky few years, during which the local management and the leadership of the Shatura Peat Trust had struggled to keep the enterprise economically viable. Pustoshinskoe's workshops had already begun to provide metal pieces for the Ministry of Light Industry in 1960. Redeployment of workers and machinery helped the trust to increase its cash flow without compromising the economic targets in its core area. Contemporaries would not have found anything unusual in this. It was common for Soviet companies to try to enhance their revenue by changing product designs or creating entirely new types of products. As producers had some influence over the pricing of newly introduced or modified produce, they could sometimes covertly improve their financial situation by inflating the price.[21] Shaturtorf's managers were not always successful when deploying this strategy. In the spring of 1967, Pustoshinskoe's director I. S. Krasnov was busy writing to ministries and planning organizations at both republican and union levels. The director was frustrated because Pustoshinskoe did not receive sufficient aluminum to produce the metal containers it was supposed to deliver to the Ministry of Light Industry. After some back and forth, his request for aluminum was denied because Pustoshinskoe's containers were not an official position in the central plan but an invention by Shaturtorf, which left the trust with little leverage to claim resources. The quick fix for this dilemma was as easy as it was cynical. Glavtorf, the central peat administration, adjusted Shaturtorf's plan so that Pustoshinskoe's actual production matched the targets.[22] The decision certainly liberated Krasnov from the pressure of having to deliver goods without sufficient supplies. But it was a solution that worked only on paper.

[20] "O peredache Pustoshinskogo torfopredpriiatiia Shaturskogo torfotresta iz vedeniia Ministerstva toplivnoi promyshlennosti RSFSR v vedenie Ministerstva khimicheskogo i neftianogo mashinostroeniia (1968)," GARF f. 5446, op. 102, d. 595, ll. 1–6; "Perepiska s torfotrestami ob izmenenii plana po proizvodstvu, trudu i sebestoimosti produktsii na 1968 god," GARF f. A-632, op. 2, d. 213, ll. 14–15.

[21] Mark Harrison and Byung-Yeon Kim, "Plans, Prices and Corruption: The Soviet Firm under Partial Centralization, 1930 to 1990," *The Journal of Economic History* 66, no. 1 (2006): 1–41, 7.

[22] "Perepiska s torfotrestami . . .," GARF f. A-632, op. 2, d. 116, ll. 184–197.

While Pustoshinskoe's workshops did not get any busier, the Ministry of Light Industry would have to look for its containers elsewhere.

In other cases, goods without the slightest link to the peat fuel industry became an important part of what peat companies produced in the late Soviet period. Satisfying the population's desire for consumer goods was a declared political aim in the decades following Stalin's death. Apartments, processed food in sterile and durable packaging, electric appliances, and clothing all entered Soviet visions of a good life. Particularly in the late 1960s and the early 1970s, consumer goods became more widely available. Even though quantity, quality, and unequal access left many citizens unsatisfied, material culture changed notably in the late Soviet period as people began to own more things.[23] Unlikely as it may seem, the peat industry was directly involved in this trend. Suffering from decades of neglect, the light industry had limited capacities to accommodate the shifting economic priorities. To implement the aim of delivering more consumer goods, the Soviet administration tried to mobilize existing and underused production facilities. From the late 1960s, peat trusts, which completed most of their core tasks during the summer, regularly received annual production targets for consumer goods (*tovary narodnogo potrebleniia*), including items needed by the light industry.[24]

On the ground, this was quite a remarkable change. Besides extracting, loading, and transporting peat, Shaturtorf's companies now regularly produced spare parts for tractors, storage boxes, metal sieves, small parts for electric items, and packaging materials. Other trusts were asked to deliver axes, wooden legs for chairs and tables, coils for soft furniture, peat-based potting soil and fertilizers, small greenhouses, as well as details for refrigerators, alarm clocks, and TVs.[25] Often struggling to meet the

[23] Chernyshova, *Soviet Consumer Culture*; Elena Kochetkova, "Milk and Milk Packaging in the Soviet Union," *Russian History* 46, no. 1 (2019): 29–52; Jenny Leigh Smith, *Works in Progress: Plans and Realities on Soviet Farms, 1930–1963* (New Haven: Yale University Press, 2014), 151–187.

[24] "O piatiletnem plane proizvodstva, truda i pribyli po Glavtorfu na 1971–1975 gody" (1971), GARF f. A-632, op. 2, d. 390, ll. 131–140; "Ob ustanovlenii plana proizvodstva i postavki tovarov narodnogo potrebleniia torgovym organizatsiiam na 1976 god," GARF A-632, op. 9, d. 53, ll. 112–113.

[25] Shatura Peat Trust, Annual Report (1976), GARF f. A-632, op. 9, d. 97, ll. 49–50; Shatura Peat Trust, Annual Report (1969), GARF f. A-632, op. 2, d. 324, l. 42; "Ob uvelichenii vypuska tovarov narodnogo potrebleniia na predpriiatiiakh Glavtorfa i uluchshenii ikh kachestva" (1971), GARF f. A-632, op. 2, d. 44, ll. 37–39; "Plan proizvodstva komplektuiushchikh izdelii" (1981), GARF f. A-632, op. 9, d. 298, ll. 223–226; "Plan proizvodstva komplektuiushchikh izdelii" (1982), GARF f. A-632, op. 9, d. 331, ll. 176–179.

government's ambitious targets, the peat industry contributed only marginally to the effort to satisfy the population's desire for modern objects. As the central peat administration pointed out in 1975, many goods prepared in the workshops of peat companies "do not meet aesthetic requirements, have low consumption value, and are not in demand among the population."[26] Still, this non-peat-based side production became embedded in the regular operations of peat trusts, compensating, at least to an extent, for the fact that peat itself had lost much of its relevance as a fuel. In 1986, when Shaturtorf's companies were making huge losses, producing spare parts, disks, carts, and other industrial goods generated at least some revenue, but not enough for the trust to break even.[27]

Mentioned in the economic record only in passing, there were also ecological limits to growth. The fact that peat companies struggled to keep afloat while the energy system underwent massive change clearly hinted at the constraints placed on bottom-up initiatives and creative innovation in the Soviet economy. But these struggles also had a material dimension specific to the peat industry: peat became scarce, a problem rooted in the very nature of the resource. Peat usually extends no more than a few meters into the ground, often less. Industrial extraction can thus exhaust the peat layer in a given spot within a relatively short period of time. Just consider the case of Pustoshinskoe. When central institutions decided the company's fate in 1968, they noted that its peat deposits were almost entirely depleted. There was not enough peat to allow Pustoshinskoe to focus solely on its initial task.[28] Scarcity became a local experience in other places as well. Compared to the idea that peat could fuel the economy for an indefinite duration of time, which had enthused late imperial and early Soviet advocates of peat, the horizon of opportunity had shrunk notably in the decades following the Second World War. In 1959, when Gosplan experts first evaluated the prospect of increasing the capacity of the Shatura Power Plant, they calculated that the surrounding peatlands could fuel the enlarged plant for 30 years.[29] In 1970, however, two years before Shatura's new unit was even commissioned,

[26] "O merakh po dal'neishemu razvitiiu proizvodstva tovarov kul'turno-bytovogo naznacheniia i khoziaistvennogo obikhoda (1975)," GARF f. A-632, op. 9, d. 1, l. 146.

[27] "Otchet proizvodstvennogo ob-edineniia Shaturtorf za 1986 god," GARF f. A-632, op. 9, d. 475, l. 41.

[28] "O peredache . . .," GARF f. 5446, oop. 102, d. 213, l 3, 6.

[29] See "Materialy po rassmotreniiu proektnykh predlozhenii nastroek na Shaturskoi, Ivanovskoi i Gor'kovskoi GRĖS (1959)," GARF f. 262, op. 5, d. 8786, ll. 27–30.

contemporaries warned that local peat reserves were running low and that Shatura would soon have to count on other fuels.[30]

Local depletion of peat deposits did not necessarily imply that the industry was running out of options. Instead, it forced peat trusts into an endless extractive cycle in which they had to abandon exploited peatlands and identify new sites where they could continue their activities every few years. In some places, this dynamic had begun to unfold in the early Soviet period, when public excitement surrounding peat was at its historic peak. In 1924, VSNKh approached Gosplan, reporting that the provisional Shatura power station had begun using peat from a new deposit because the previous site had been "long exhausted." At that time, this was considered first and foremost an economic matter. The planners needed to know how much the peat from the new site was worth in money to calculate the price for the end consumers of electricity.[31] Over the years, the gradual depletion of peatlands became a growing challenge. Peat works increasingly encroached on the edges of extraction fields, hill slopes, or poorly connected areas, where extraction took much more time and effort. The diffusion of the milled-peat extraction method from the late 1950s initially helped hide that mobilizing peat resources for Russia's fossil economy was becoming increasingly costly and onerous. However, administrators noted in 1967 that the hopes for productivity increases linked to the new machinery did not always materialize, as peat companies turned to less suitable deposits.[32] In the 1980s, some peat companies even began exploiting so-called unbalanced (*zabalansovye*) peat reserves, which were not formally considered suitable for extraction because they were difficult to access or contained peat of low quality.[33] This strategy may appear to be a typical example of Soviet resource mismanagement, but it was not exceptional. As the geographer Gavin Bridge put it, "the production of more from less (i.e., from lower quality stocks)" was in many parts of the world a crucial feature of how resources were appropriated and mobilized during the twentieth century.[34]

30 Lipenskii, *Bol'shaia Shatura*, 180.

31 VSNKh to Gosplan (January 12, 1924), RGAĖ f. 4377, op. 5, d. 255, l. 4.

32 "Doklad Glavtorfa o sostoianii tekhnicheskogo urovnia proizvodstva v torfianoi otrasli za 1967 god," GARF f. A-632, op. 2, d. 134, l. 15.

33 "Poiasnitel'nia zapiska po Iaroslavskomu proizvodstvennomu ob-edineniiu po dobyche torfa" (1984), GARF f. A-632, op. 9, d. 402, l. 34; "Poiasnitel'naia zapiska po Iaroslavskomu proizvodstvennomu ob-edineniiu po dobyche torfa" (1986), GARF A-632, op. 9, d. 455, l. 35.

34 Bridge, "Material Worlds," 1225.

The official response to these problems was not to end or reduce peat extraction but to amend its purpose and geography. Experts and administrators remained committed to the extractive use of peatlands. Their rhetoric, however, changed considerably, reflecting an increasing concern about what late Soviet planners and technical elites began to call the "rational utilization" (*ratsional'noe izpol'zovanie*) of natural resources and the changing economic priorities articulated by the political leadership. In 1978, the Ministry of Geology warned the Council of Ministers of the RSFSR that long-term exploitation had severely reduced peat reserves in central Russia. Under the current extraction levels, the remaining deposits would last no longer than 5–10 years in the region of Moscow.[35] Fedor M. Morozov, the republic's Deputy Minister of Geology, and Vitalii V. Bogorodskii from the Academy of Sciences elaborated on this point in a prominently placed *Pravda* article the following year. The scarcity of peat in central Russia, they argued, only highlighted how irresponsibly it was being utilized: Burning peat in power plants and furnaces was a wasteful practice and ignorant of other possible applications for peat. Using it as a growing medium, fertilizer amendment, insulation material, or as a source of ammonium and other valuable substances now seemed the imperative of the hour. The rationale underpinning their argument was hard to miss. Given that new energy carriers were becoming increasingly accessible, while Soviet agriculture and chemistry remained underdeveloped, peat lost its earlier appeal as a fuel source. One passage of the article, which was reminiscent of earlier Cornucopianism, stressed that, in certain regions, such as the Russian northwest, western Siberia, the Urals, Kamchatka, and Sakhalin, peat resources were large enough to satisfy the needs of the national economy "for a long time" (*na dlitel'nyi srok*).[36] After the peat industry had literally undermined the means of its own existence in central Russia, the extractive frontier was bound to move on.[37]

[35] "O ratsional'nom ispol'zovanii zapasov torfa v Nechernozemnoi zone RSFSR (1978)," GARF A-262, op. 5, d. 5170, ll. 28–29.

[36] F. Morozov and V. Bogorodskii, "Obyknovennyi torf," *Pravda* (June 4, 1979): 2.

[37] This geographic shift of the peat industry is also described in A. L. Yampolsky, "Methodological Principles of Projecting the Development and Distribution of the Peat Industry in the USSR," *VIII International Peat Congress. Proceedings. Vol. II*, Leningrad, 1988, 23–27. Post-Stalinist Soviet forestry followed a similar approach, expanding into Siberia and the Far East to address the overexploitation of forests to the West of the Ural mountain chain. See Kochetkova, *Green Power of Socialism*, 29–67.

Published half a year after the government had announced the phasing out of peat fuel, the article reflected the changing parameters under which the peat industry operated in the late Soviet period. When economic planners set the sector's production targets for the period 1981 to 1985, the picture looked a lot different from what it had been when the Soviet state had initially set up its peat-based power infrastructure. The planners were still committed to the growth paradigm, projecting an overall 5.6 percent increase in peat extraction for the RSFSR. The spatial organization of extraction and the allocation of peat in the broader economy, however, had begun to shift. Not only did the plan project that extraction for fuel would decline in absolute terms while agricultural peat would gain in relative and absolute importance. It also determined that extraction would decrease in many regions that had once spearheaded the economic mobilization of peatlands. After decades of serving the Soviet power industry, the central Russian peat trusts Shatura, Vladimir, Orekhovo-Zuevo, Yaroslavl, and Gorky were all projected to reduce their output. Shatura, whose activities had already slowed, was supposed to further reduce its fuel-peat extraction by a quarter. The peat trusts Vologda, Leningrad, Pskov, Smolensk, Sverdlovsk, and Tyumen would, instead, either maintain or expand their output levels.[38] In the early 1980s, Russia's peatlands were still regarded as a resource, though the parameters and aims of extraction had taken on a new guise.

EXPLOITED PEATLANDS AS HORIZONTAL RESOURCES

The fact that the peat industry moved slowly but inexorably across the territory raised important questions as to the environments it left behind. What was supposed to happen to a peat extraction site after its reserves were depleted? Shifting from a vertical to a horizontal perspective, scientific experts and planners were optimistic that exploited peatlands had yet to exhaust their potential as resources: drained, cleared, and stripped of most of their peat, these lands could become farmland or forests. This idea had initially gained shape during the late imperial period, even though it did not feature prominently in the academic and public debates about peat fuel at the time; that the depletion of peat would ever become a problem did not seem plausible as long as peat was believed to exist in

[38] "Plan po dobyche torfa na 1981-1985 gody po proizvodstvennym ob-edineniiam Rostorfa," GARF f. A-632, op. 9, d. 298, ll. 131–132.

endless quantity.[39] With the rise of large-scale industrial exploitation and notably the transition to milled peat extraction in the second half of the twentieth century, the energy hunger of Russia's fossil economy touched ever more extensive wetland areas. As a result, concerns about what would follow extraction became ever more urgent. The conversion of former extraction sites into spaces for food production or industrial forestry was generally embraced as the most attractive option. Where the peat industry left quarries filled with water (as was the case for older extraction methods), they were considered suitable for fisheries. From this point of view, peat extraction appeared as one stage on a pathway which would render peatlands "productive landscapes" in the long run.[40]

In practice, peat extraction left a trail of destruction. Preoccupied with production targets, neither the directors of peat trusts nor the peat workers had much capacity to place any thought or effort into preparing peatlands for productive after-use. The authorities struggled to address the problem. P. V. Bibikov, Gosplan expert on peat matters since the early years of Soviet power, lamented in 1944 that many companies exploited peatlands in a predatory way without bothering to leave the land in such a state that it could be used as farmland. For Bibikov, this neglectful approach to resources had been embedded in the history of Soviet peat extraction from early on: "We've been talking about this for thirty years, but they extract just as they want."[41] The following year, the authorities agreed on rules to tackle the issue. Depending on the extraction method, peat companies were supposed to leave a peat layer of 10–15 centimeters to facilitate plant growth once extraction ended. They also had to respect certain norms when designing drainage systems and were asked to keep the edges of extraction sites straight.[42] What these regulations did not address, however, was the fact that peat trusts had no reason to care about peatlands after they had exploited them, while the state had little to no leverage to enforce these rules. Once they were finished, the enterprises

[39] That exploited peatlands were seen as future farmland as early as imperial times shows in Vikhliaev's critical remarks on peasants' careless use of peatlands during the First World War. See *Trudy soveshchanii po podmoskovnomu ugliu i torfu*, 89.

[40] V. N. Sokolov, *Torf v narodnom khoziaistve* (Moscow: Nedra, 1988), 256; V. D. Markov, "Ispol'zovanie vyrabotannykh torfianykh kar-erov," in A. A. Nitsenko and M. S. Boch, eds., *Priroda bolot i metody ikh issledovanii* (Leningrad: Nauka, 1967), 233–236.

[41] "Stenogrammy zasedanii mezhvedomstvennogo tekhnicheskogo Soveta pri Glavtorffonde NKZ RSFSR" (1944), GARF f. A-412, op. 1, d. 2, l. 8.

[42] "Pravila i instruktsii po kontroliu za izspol'zovaniem torfianykh mestorozhdenii (1945)," GARF f. A-412, op. 1, d. 13, l. 84, ll. 145–189.

often did not even hand the land over to the authorities so that they could ensure further exploitation. As the Chief Administration of the Peat Fund noted in 1945, many former peat extraction sites in the region of Moscow were "abandoned degraded lands" (*bezprizornye brosovye zemli*).[43]

Criticism of the peat industry and its destructive impact on the landscape took on a new form during the years of the Thaw when it became part of a broader conversation about the negative consequences of extractivism and growth. In the late 1950s, scientists and writers took advantage of the newly emerging opportunities for public expression to demand that environmental principles be anchored in Soviet law and institutions. Moving away from a narrow focus on the protection of selected places or landscapes, which had characterized the preservation movement since the late imperial period, they articulated a more extensive understanding of nature protection, demanding that the state should play a more active role in tackling human-made problems such as pollution and resource overuse. The Academy of Science, which developed the most ambitious project of this period, advocated for a central organ committed to protecting nature and a corresponding law that would be put in force in all Soviet republics.[44] During their attempts to solicit support for their idea, Alexander Nesmeianov, the President of the Academy, and K. Segal, a member of the Academy's Nature Protection Commission, warned that the Soviet Union was about to confront a scenario which was known so far only from the capitalist world, "the deep and irreversible depletion of natural resources."[45] To underscore the urgency of the matter, the scientists compiled a comprehensive list of environmental problems facing the country. The condition of abandoned peat extraction sites was one of the many reasons why they demanded state-led nature protection efforts on a comprehensive scale. Industrially exploited peatlands bore little resemblance to successfully reclaimed land. The Academy's researchers found that most peatlands were "in an ugly state," hardly capable of sustaining any noteworthy plant growth, and "almost like a desert."[46] The finding must have had a bitter taste. Rather than eliminating wastelands, the peat industry had actually created them.

[43] "Otchet o vypolnenii plana rabot po Gosudarstvennomu Torfianomu Fondu za 1946 po Moskovskoi oblasti," GARF f. 412, op. 1, d. 28, l. 209.

[44] Coumel, "A Failed Environmental Turn?"

[45] "Predlozheniia komissii po okhrane prirody AN SSSR v Gosplan SSSR 'O meropriiatiiakh po sokhraneniiu i vosstanovleniiu prirodnykh resursov' dlia vkliucheniia v narkhozplan 1959-1965gg.," RGAĖ f. 544, op. 1, d. 60, l. 3.

[46] Ibid., l. 109.

The most obvious solution to the problem, stopping extraction altogether, was not considered an option. Peat was still so embedded in the Soviet energy system that few people at the time would have questioned the existence of the peat industry. On the contrary, suggestions to address the dreadful state of exploited peatlands followed a line of reasoning which was widely endorsed in late Soviet environmental thought. Socialism's ability to destroy nature might have been similar to that of capitalism, but socialist society was much better equipped to address and prevent this problem. Science, planning, and state regulation would balance human activity with the capacity of the natural world without sacrificing the promise of comfort and affluence that Soviet leaders made to the population.[47] An influential expression of this technocratic optimism was the concept of "rational" or "complex utilization of natural resources," which became a major framework for environmental science and policy in the late Soviet period. Recognizing that economic expansion involved costs that threatened to undermine the prospects of future growth, rational resource use implied that, if appropriately managed and controlled, extractive industries could avoid pollution and wastefulness.[48]

The 1960 Russian Nature Protection Law introduced this new language to the legal sphere. Its scope and geographical reach paled compared to what the Academy's scientists had envisaged, but the legislation nonetheless echoed their concerns about degraded industrial sites. The law testified to the belief in regulation and monitoring as means to ensure that extractive industries would treat natural resources with care. Determining that mineral-extracting enterprises (peat extraction was explicitly included in this category) were under a legal obligation "to carry out measures to restore the fertility of lands affected by their project and suitable for agricultural use," the law created a frame of reference for experts and planners who would try to make the peat industry more considerate of the limits to growth and the dangers of environmental degradation.[49]

[47] Soviet cetologists who pioneered scientific research on whale communication and behavior saw science as a means to manage whale populations. Jones, *Red Leviathan*, 134–150. Ideas of technocratic resource management also informed the Soviet forestry industry. Kochetkova, *Green Power of Socialism*.

[48] Andy Bruno traced this concept back to the Stalinist period. Bruno, *Nature of Soviet Power*, 97. For a transnational perspective on the idea of "integrated" resource use, see Coumel, "Soviet Eco-Power," 18–21.

[49] The translation is from Marshall I. Goldman, *The Spoils of Progress: Environmental Pollution in the Soviet Union* (Cambridge MA: MIT Press, 1972), 303. For the original, see "Ob okhrane prirody v RSFSR," *Pravda* (October 28, 1960): 2.

By no means did the inclusion of abandoned peat extraction sites into environmental rhetoric and policies end the productivist approach to wetlands that had dominated elite discourse since the Enlightenment era. Instead, the concern for the poor state of peatlands translated into another type of productivism, one which projected ideas of agricultural improvement onto lands that had been industrially exploited. Environmentalist rhetoric in this context functioned as a device not to protect peatlands from the logic of economic expansionism but to maintain them within the resource pool of the Soviet economy. The keyword marking the rising awareness of the destructive impact of industrial extraction and the widespread belief in the possibility of rectifying these problems was "recultivation" (*rekul'tivatsiia*), which became an important area of scholarship and a policy framework following the adoption of the Nature Protection Law. In the case of peatlands, recultivation implied the appropriation of abandoned peat deposits for farming, forestry, or fisheries.[50] Considering the changing role of peat in Russia's fossil economy, it is hard not to see an irony in the rhetoric surrounding the concept. At the same time as peat-extracting enterprises and administrators in government agencies lamented that peat was becoming scarce, exploited peatlands invoked new visions of abundance and plenty. Recultivation, as one publication put it, promised peatlands "a second life," reinvigorating the "barren wastelands" left behind by extractive exploitation.[51] This approach to undoing industrial damage also aligned with the massive land reclamation programs in the late Soviet period, when millions of hectares of land became subject to irrigation and drainage. In a comprehensive study on the intersection of land reclamation and nature protection, drainage experts Boris S. Maslov and Ivan V. Minaev called exploited peatlands a "significant reserve for the expansion of agricultural and forest land."[52]

In contrast to the hope and optimism that the prospect of cultivating exploited peatlands instilled, reality looked rather bleak. In most places, converting exploited peatlands into bountiful fields or productive forests remained no more than an aspiration. Recultivation never caught up with the scale and speed of the peat industry. In 1976, the Soviet government

50 "Otchet o nauchno-issledovatel'skoi rabote Tsentral'noi Laboratorii okhrany prirody za 1965 god," RGAĖ f. 7486, op. 33, d. 23, ll.1–51.

51 Iu. V. Zverkov, *Vtoraia zhizn' torfianikov* (Kirov: Volgo-Viatskoe kn. izd-vo, 1982), 7.

52 B. S. Maslov and I. V. Minaev, *Melioratsiia i okhrana prirody* (Moscow: Rossel'khozizdat, 1985), 11.

obliged extracting enterprises to ensure that the land they used remained in a state which allowed a smooth transition to productive after-use. If their actions had in any way harmed the fertility of the soil, enterprises now had to undertake restoration works.[53] The Academy of Sciences had noted a few years before that recultivation requirements were met more often in exploited peatlands than in territories mined for coal or minerals or places that were otherwise polluted by resource extraction.[54] In the Shatura region, for instance, the Sovkhoz farm *Shaturskii* successfully grew potatoes and cabbage on exploited peatlands as early as 1963.[55] Still, large areas of peatland were simply abandoned, once most of their peat was gone. By the late 1970s, no more than a third of all exploited peatlands in the Soviet Union had been returned to the previous landowners and, of these, only half were actually put to use. Left to themselves, exploited peatlands either reverted to a swampy state (*zabolachivanie*) or were exposed to wind and precipitation, which meant that the remaining peat layer was effectively lost.[56]

Besides peat trusts' limited willingness or preparedness to engage in recultivation, the key reason for the ever-growing number and size of degraded peatlands was the expansive nature of the peat industry. Even when peat-extracting enterprises incorporated recultivation into their operations, this did not mean that their overall imprint on the land got any smaller. Instead, they kept encroaching on previously unexploited wetlands. This was, after all, precisely what the growth-oriented economic administration expected from them. Just in the second half of 1978, peat trusts in the RSFSR were supposed to undertake extraction in more than 11,000 hectares of previously unexploited peatlands. At times, the commissioning of new sites was explicitly presented as compensation for the renunciation of exploited deposits.[57] The problem

53 "Postanovlenie Soveta Ministrob SSSR o rekul'tivatsii zemel', sokhranenii i ratsional'nom ispol'zovanii plodorodnogo sloia pochvy pri razrabotke mestorozhdenii poleznykh iskopoaemykh i torfa, provedenii geologorazvedochnykh , stroitel'nykh i drugikh rabot (June 2, 1976)," in *Resheniia partii i pravitel'stva po khoziaistvennym voprosam: Sbornik dokumentov. Tom 11* (Moscow: Politizdat, 1977), 337–341.

54 "Ob uluchshenii organizatsii rabot po rekul'tivatsii zemel', narushennykh i zagriaznennykh promyshlennost'iu v SSSR (1966)," RGAĖ f. 544, op. 1, d. 176, l. 9.

55 L. A. Gorshkov, et al., *Kompleksnoe ispol'zovanie torfianykh bolot* (Moscow: Rossel'khozizdat, 1969), 147.

56 Zverkov, *Vtoraia zhizn' torfianikov*, 7.

57 "Zadanie po podgotovke novykh proizvodstvennykh ploshchadei dlia dobychi frezernogo torfa po proizvodstvennym ob-edineniiam Rostorfa na vtoroe polugodie 1978 goda," GARF, f. A-632, op. 9, d. 153, l. 102; "Zadanie po pogdotovke novykh proizvodstvennykh

only increased as the peat industry in central Russia began to decline. In 1986, the Ministry for Land Reclamation warned that thousands of hectares of exploited peatlands remained unused in the Meshchera Lowland.[58] Contemporary experts feared the long-term damage this might cause, but there was little hope that this issue would ever be addressed systematically. At a time of deepening economic crisis, farming, forestry, and fisheries could not expand their activities at the same pace as exploited peatlands became available. As a result, many degraded peatlands were left to themselves. In practical terms, this was a radical departure from the productivist imperative previously imposed on wetland ecosystems. The nature of this departure, however, created novel forms of ecological precarity whose full extent would become evident in the new millennium.

TOWARD A NEW WETLAND PARADIGM

While most attempts to prevent or rectify the damage caused by the peat industry revolved around productivist land imaginaries, the late Soviet period also saw the rise of a new understanding of wetlands, which pointed to their role as habitats, water regulators, and recreational spaces. From this point of view, peat extraction appeared in an entirely new light: The diggers and milling machines which sliced their way through central Russia's peatlands were not harbingers of cheap energy or reclaimed land but agents of violent and irreversible destruction. Such ideas surfaced when the scale of degradation became increasingly apparent. At the same time, they gained traction as part of a transnational trend, during which the deteriorating state of wetlands attracted the attention of conservationists from east and west of the Iron Curtain and the Global South. In 1971, the Soviet Union was among the signatories of the Ramsar Convention on Wetlands of International Importance, which defined wetlands as an object of environmental law and underscored the need to prevent their loss. Realized under the umbrella of UNESCO, the convention marked the consolidation of a new paradigm and has since informed wetland protection measures

ploshchadei dlia kompensatsii vybyvaiushchikh moshchnostei po proizvodstvennym obedineniiam Rostorfa na 1983 god," GARF A-632, op. 9, d. 331, l. 79.

58 "O merakh po okhrane prirody i ėffektivnosti meliorativnykh rabot organizatsii Glavnechernozemvodstroia v Meshcherskoi nizmennosti (1986)," GARF A-259, op. 48, d. 9284, l. 35.

worldwide.[59] It was within this wider context that advocacy for peatlands grew louder within the Soviet Union.

Wetlands did not have to be destroyed for their importance to be acknowledged though. By the time Soviet scientists, artists, and ordinary citizens began to show concern for degraded peatlands, the distinct features and fragile nature of these ecosystems were long known. Community ecology, an intellectual tradition that originated in the imperial and thrived in the early Soviet period, produced a substantial body of scholarship on plant and animal life and the complex agencies of water and matter in peatlands. Community ecologists highlighted the interaction among different species and the effects of geological and climatic factors on living organisms. Their aspiration to theorize the relationship between living and nonliving matter within specific locations culminated in the term *biogeocenos,* which was coined by geobotanist Vladimir N. Sukachev in the 1940s and soon became a key concept in Soviet sciences. Crucially, Sukachev's early work was linked to the tradition of *bolotovedenie*, the study of wetlands which, like the broader field of community ecology, had roots in the late imperial period. Published first in 1914 and re-edited several times during the Soviet period, Sukachev's *Swamps: Their Formation, Development, and Properties* was a landmark study in what would nowadays be called wetland ecology.[60] In this book, Sukachev explained how vegetation, precipitation, soil conditions, and geography coalesced in the formation of different types of peat and peatlands, and emphasized how studying these landscapes could inform disciplines as diverse as geology, soil science, climatology, botany, geography, and archaeology.[61] Sukachev's fascination with peatlands significantly influenced the activities of the Academy of Sciences, particularly

[59] Edward Maltby, "The Changing Wetland Paradigm," in Edward Maltby and Tom Barker, eds., *The Wetlands Handbook* (Chichester: Wiley-Blackwell, 2009), 3–42; "International Conference on the Conservation of Wetlands and Waterfowl: Convention on Wetlands of International Importance," *International Legal Materials* 11, no. 5 (1972): 963–976. Parts of this section contain rewritten passages from my chapter Katja Bruisch, "How Wetlands Entered the Transnational Spaces of Late Soviet Environmentalism," in Catherine Evtuhov, Julia Lajus, and David Moon, eds., *Thinking Russia's History Environmentally* (New York: Berghahn, 2023), 201–221. I thank Berghahn for the permission to reuse this material.

[60] Douglas R. Weiner, *Models of Nature: Ecology, Conservation and Cultural Revolution in Soviet Russia. With a New Afterward* (Pittsburgh: University of Pittsburgh Press, 2000), 23, 64–65.

[61] V. N. Sukachev, *Bolota: Ikh obrazovanie, razvitie i svoistva.* Second edition (Petrograd: Novaia Derevnia, 1923), 5.

once it became more invested in studying the human impact on the environment. In 1969, a new research unit dedicated to studying "peatland biogeocenoses" (*po izucheniiu biogeotsenozov bolot*) was established within the Academy's Scientific Council for Problems of Biogeocenology and Nature Protection.[62]

Since productivist wetland imaginaries persisted in the Soviet Union, peatland conservation remained a difficult sell. Despite its influence as a scientific framework, there was no direct line running from community ecology to later demands for the active protection of peatlands. Mirroring the heterogeneous nature of late Soviet environmental thought, the Academy's engagement with peatlands oscillated between the productivist ideal of rational resource use and a more explicit commitment to preventing wetland intrusive uses. Nikolai I. Piavchenko, once a student of Sukachev's and the head of the new research unit, embraced the former approach, providing arguments and expertise in favor of drainage reclamation to support tree growth.[63] In spite of, or rather responding to, the drainage frenzy of the Brezhnev period, some of Piavchenko's colleagues actively advocated peatland protection. These scholars benefited from the rise of environmental diplomacy during the Cold War when the Soviet leadership was attracted by "the appeal of appearing green," as the historian Stephen Brain put it, and supported Soviet participation in relevant international negotiations and research initiatives.[64]

Founded upon a joint initiative by the International Union for Conservation of Nature and the International Biological Program in 1966, the Telma project constituted one of the newly emerging spaces in which Soviet wetland scientists collaborated with colleagues from outside of the Soviet bloc. Aimed at protecting peatlands as sites of scientific inquiry, Telma (from the Greek *telma* for "mud") was a reaction to the alarming condition of peatlands due to farming, forestry, and peat

62 "O rabote podsektsii po izucheniiu biogeotsenozov bolot Nauchnogo Soveta Akademii Nauk SSSR po probleme 'Kompleksnoe biogeotsenologicheskoe izuchenie zhivoi prirody i nauchnye osnovy ee ratsional'nogo osvoeniia i okhrany," ARAN, f. 2067, op. 1, d. 9, ll. 12–14.

63 N. I. P'iavchenko, ed., *Vzaimootnosheniia lesa i bolota: Po dannym statsionarnykh issledovanii* (Moscow: Nauka, 1967); "Tezisy vsystupleniia na temu 'Znachenie biogeotsenologii v osushitel'noi melioratsii', N. I. P'iavchenko," ARAN f. 2067, op. 1, d. 9, l. 38.

64 Stephen Brain, "The Appeal of Appearing Green: Soviet-American Ideological Competition and Cold War Environmental Diplomacy," *Cold War History* 16, no. 4 (2016): 443–462.

extraction in many parts of the world.[65] Under the leadership of geobotanist Andrei A. Nitsenko, Piavchenko's Deputy in the Academy's peatland unit, the project gained a foothold in the Soviet Union. Members of the Telma group compiled lists of important peatlands and raised the need to protect them with Soviet planning bodies. Even though they could draw on a rich domestic tradition of peatland science, the scholars used their involvement with a transnational initiative to leverage their concerns domestically. Speaking to the All-Union Meeting on Questions of Peatland Protection (*po voprosam okhrany bolot*) in 1974, scientist Viktor V. Mazing, the new chair of the Soviet Telma group, expressed his delight that a novel understanding of wetlands had gained ground in the Soviet Union. Mazing also urged that concrete measures for protection be undertaken locally. Several small achievements seemed to justify his optimism. Botanist and member of the Soviet Telma group Marina S. Botch reported at the same event that Gosplan, the Soviet Ministry of Agriculture, RSFSR's Ministry of Forestry, and the Russian Peat Fund had all agreed to abstain from exploiting several peatlands which had been identified as particularly valuable.[66]

While government agencies continued to promote drainage reclamation, Soviet wetland science increasingly gravitated away from productivist concepts of nature. During the 1970s, several scholars distanced themselves from the idea that their research should be guided by the interests of farming, forestry, or the peat industry. Instead, they underscored what made intact peatlands so valuable and preserving them so urgent. Important markers of this trend were Konstantin E. Ivanov's *Water Movement in Mirelands* (1975, translated into English in 1981) and *Mire Ecosystems in the USSR* (1979), co-authored by Marina Boch and Viktor Mazing.[67] Proposing an ecological rather than economic approach to determining what nature was worth, both publications stressed peatlands' role as habitat and regulators of water flows. In contrast to the land reclamation discourse that focused on drainage as an instrument to render useless wetlands useful, these authors pointed

[65] David J. Bellamy and Tom Pritchard, "Project 'Telma': A Scientific Framework for Conserving the World's Peatlands," *Biological Conservation* 5, no. 1 (1973): 33–40.

[66] M. S. Boch, "Vsesoiuznoe soveshchanie po voprosam okhrany bolot (Leningrad, 15 maia 1974g.)," *Botanicheskii zhurnal* 61, no. 2 (1976): 291–295.

[67] K. E. Ivanov, *Vodoobmen v bolotnykh landshaftakh* (Leningrad: Gidroprometeoizdat, 1975); M. S. Boch and V. V. Mazing, *Ėkosistemy bolot SSSR* (Leningrad: Nauka, 1979). I have used "mire" in the translation of these titles, for this is the term the scientists would use in most of their publications in English.

to their fragility. Turning long-established notions of peatlands onto their head, Nitsenko and Boch argued in one publication that most human interference was not "correcting" but rather damaging. Peatlands, they wrote, "cannot be economically used without being destroyed as such." Drainage and industrial exploitation robbed peatlands of precisely the features which made them distinct. Counteracting these trends was an urgent matter, particularly in regions where extraction and reclamation progressed rapidly: "soon nothing will be left to preserve."[68]

Pushing the conventions and boundaries of Soviet wetland sciences, this understanding opened the field up to concerns about ecological vulnerability and loss that also informed the rise of the new wetland paradigm abroad.[69] Seeing themselves as part of a global conversation, some Soviet scientists argued that they should actively develop a language allowing them to share their insights across political and linguistic borders. In his opening contribution to the comprehensive volume *Mire Types in the USSR and Principles of Their Classification*, published by the Academy of Sciences in 1974, Mazing wrote:

> The Soviet study of mires and peatlands (*bolotovedenie*) has contributed quite substantially to the treasury of world science. However, the fact that we use our own terminology (*mestnaia terminologiia*) and care little to make our work internationally known, read, and understood prevents us from taking our well-deserved place in the avant-garde of this discipline. Respecting our accomplishments, traditions, and language, we do also have to consider the internationally most common terminological conventions and make sure that our contributions will be acknowledged at the international level.[70]

Adopting internationally accepted terminology, Soviet scientists tried to attune the Soviet discourse to present challenges. When Mazing and Botch published their *Mire Ecosystems of the USSR*, they were conscious that the term "ecosystem" in the title challenged the undisputed position of Sukachev's "biogeocenosis" concept, which had previously dominated the field. The authors were careful to affirm that "biogeocenosis" could

[68] M. S. Boch and A. A. Nitsenko, "Ob okhrane bolot v SSSR," in A. M. Semenova-Tian-Shanskaia and N. A. Frantskevich, eds., *Voprosy okhrany botanicheskikh ob-ektov* (Leningrad: Nauka, 1971), 36–37, 39.

[69] Anna-Katharina Wöbse, "Vom Ödland zum Reservoir des Reichtums: Die Neudeutung europäischer Sümpfe," *Ilinx—Berliner Beiträge zur Kulturwissenschaft* 5 (2020), 113–125.

[70] V. V. Mazing, "Aktual'nye problemy klassifikatsii i terminologii v bolotovedenii," in T. G. Abramova, M. S. Boch, and E. A. Galkina, eds., *Tipy bolot SSSR i printsipy ikh klassifikatsii* (Leningrad: Nauka, 1974), 11.

be applied to specific mires or peatlands. However, they considered the broader "ecosystem" better suited for the systematic study they had undertaken because it referred to a type rather than a concrete ecological formation.[71] But the term held other advantages, too. Allowing Soviet scientists to link themselves with the international scientific community on the one hand, it opened up new horizons to leverage peatlands as a category of nature protection domestically on the other.

Scientists were not alone in raising the alarm about the worrying state of wetlands. In the late Soviet period, writers played a crucial role in bringing environmental issues to the attention of a wider public. Their texts were often imbued with nationalist rhetoric that, like the literary production of the late nineteenth century, suggested a direct link between nature and Russian culture.[72] The central role of literary journals as platforms for concerns about humans' destructive power is evident in the case of wetlands. In 1965, *Novyi Mir* published an extensive essay by widely known village-prose author Gavrill N. Troepol'skii, who exposed predatory resource use, the wasteful spending of public money on useless environmental engineering schemes, and the negative impacts of land drainage such as river shallowing, soil erosion, and the drying up of vegetation.[73] Although articles like his did not interrupt any ongoing drainage schemes, they certainly contributed to the growing public uneasiness about large-scale interventions in wetland ecosystems. In 1976, literary magazine *Literaturnaia Gazeta* featured an article in which the writer Iurii P. Vronskii pointed out that reclamation inflicted severe harm on peatlands, for instance by causing the loss of moorberry habitat.[74] Drainage skepticism gained such momentum in the following years that, in 1983, *Literaturnaia Gazeta* invited land reclamation expert Boris S. Maslov to address readers' anxieties about the negative impact of drainage. Maslov, who would champion land reclamation until he died in 2015, used the opportunity to stress that the designers of most Soviet drainage projects carefully considered potential ecological damage. For him, drainage was still the best way to develop marshy pieces of land.[75]

71 Boch and Mazing, *Ėkosistemy bolot*, 6.

72 See, for instance, Breyfogle, "At the Watershed"; N. M. Dronin and J. M. Francis, "Econationalism in Soviet Literature," *Soviet and Post-Soviet Review* 45 (2018): 51–72.

73 G. Troepol'skii, "O rekakh, pochvakh i prochem," *Novyi Mir* 41, no. 1 (1965): 185–212.

74 Iu. Vronskii, "Ostorozhno boloto," *Literaturnaia Gazeta*, no. 14 (1976): 2.

75 B. Maslov, "Esli v kolodtse ischezla voda . . .," *Literaturnaia Gazeta*, no. 7 (1983): 13.

The Meshchera Lowland occupied a prominent place in the public revaluation of peatlands. While Trotsky, who had written about the region a few years after the Bolshevik takeover, saw the wolves and elks near Shatura as indicators of backwardness, Meshchera's remaining wildlife came to stand for its uniqueness as a habitat as time progressed.[76] The 1964 edition of the Atlas of the Moscow region suggested the "Meshchera wetlands" (*Bolota Meshchery*) as a tourist destination. The following 1976 version highlighted their peculiar flora and fauna. The area's environments were rich in leafy forests, moorberries, and water-loving plants like menyanthes. Among the many animal species that had their home here were beavers, black grouses, capercaillies, mink, and the endangered Russian desman. Meshchera's marshy lands also attracted large numbers of migrating birds which regularly rested along the shores of the Oka River bordering the region in the South.[77]

These imaginaries were not uncontested, however. Even when peat extraction declined, Meshchera continued to evoke visions of agricultural abundance. Although acknowledging the region's rich natural heritage, planners continued seeing land drainage as a way to support the ailing farming sector. In 1986, for instance, the Ministry of Land Reclamation, the State Bank, and several regional authorities discussed the insufficient implementation of reclamation works in the Meshchera Lowland, its low level of agricultural development, and its underused soils.[78] At that moment, however, the tide was already turning. Scientists had entertained the idea of protecting parts of the lowlands since at least 1982. Plans to further exploit Meshchera's nature for the economy came under open public fire during Perestroika, when the Soviet public became more aware that the government's costly drainage operations did not yield the expected economic reward but destroyed environments that people appreciated as spaces where they could pick berries, go fishing, or for leisure and recreation. In September 1985, shortly after Mikhail Gorbachev came to power, a *Pravda* article openly criticized plans to drain the land and extract peat in the Meshchera, particularly near the Oka River. The author stressed the importance of the wetlands as habitat

[76] I discuss the myths surrounding Shatura and the Meshchera in Chapter 4.

[77] *Atlas Moskovskoi Oblasti* (Moscow: Glavnoe Upravlenie Geodezii i Kartografii, 1964), 12; *Atlas Moskovskoi Oblasti* (Moscow: Glavnoe Upravlenie Geodezii i Kartografii, 1976), 19.

[78] "O merakh po okhrane prirody i ėffektivnosti meliorativnykh rabot organizatsii Glavnechernozemvodstroia v Meshcherskoi nizmennosti (1986)," GARF A-259, op. 48, d. 9284, ll. 1–48.

and highlighted their significance for the region's larger ecological balance.[79] While the Meshchera National Park was not created until the early 1990s, the article demonstrated that a new wetland paradigm had gained solid ground in the late Soviet period.[80]

EXTRACTION'S UNCANNY AFTERLIVES

For communities, whose existence was intrinsically tied to the peat industry, the revaluation of peat and peatlands turned into a cause of severe concern during the economic reforms of the Perestroika period. In the late 1980s, the Soviet leadership under Mikhail Gorbachev experimented with deregulation to return the country to economic growth by giving enterprises more leverage over their production and strengthening principles of profit-and-loss accounting.[81] The power industry's move away from peat fuel, combined with the government's economic tinkering, caused unseen difficulties for the Shatura Peat Trust. Instead of relying on the state to cover economic losses, as had been the norm in previous decades, the trust now had to search for profitable business models to keep financially afloat even though the demand for its main product was shrinking. Sometimes, these pressures led to staff layoffs. In most cases, however, the trust's companies tried to retain workers by generating new income streams. The production of consumer goods, an activity that had long been embedded in the trust's daily operations, now became decisive to keeping peat companies viable. In 1990, when the country was mired in economic problems, Shaturtorf's newspaper *Za Torf* featured a regular column called "The new produce," reporting on how its companies introduced new lines of production to adapt to the changing realities: the peat company Tugolesskii Bor manufactured radio capacitors; the Baksheevskoe company sewed shorts and sportswear for children; others made wooden furniture, backpacks, shoes, or boots.[82] Doing all this during a severe supply crisis, when even the most basic tools and materials

[79] G. Ryzhenkov, "Molnii nad bolotom," *Pravda* (September 3, 1985): 3.

[80] Roe, *Into Russian Nature*, 227–231.

[81] Donald A. Filtzer, "The Contradictions of the Marketless Market: Self-Financing in the Soviet Industrial Enterprise, 1986–90," *Soviet Studies* 43, no. 6 (1991): 989–1009.

[82] "Trudnaia pora osvoeniia," *Za Torf* (February 2, 1990): 1; "Vklad shveinogo tsekha," *Za Torf* (February 13, 1990): 1; "Na mebel'nom uchastke," *Za Torf* (February 13, 1990): 2; "Devchatam nravitsia," *Za Torf* (April 13, 1990): 1; "V polose osvoeniia," *Za Torf* (May 22, 1990): 1; "Perekvalifikatsiia neizbezhna," *Za Torf* (June 15, 1990): 1; "Osvoili novye spetsial'nosti," *Za Torf* (December 21, 1990): 1.

were hard to get, was extremely difficult. Nevertheless, at least for a while, these activities seemed to suggest a possible avenue for the trust to navigate the changing realities.[83]

The growing emphasis on profitability meant that the dual role of the enterprise as an economic and a social unit was called into question. The signs were small at first. In mid-1990, Shaturtorf's leadership decided that the publication of the newspaper *Za Torf*, which sold for the symbolic price of one kopek per issue, was too costly to be continued in the following year. After an outpouring of anger and grief, the implementation of the decision was delayed. Still, it was a clear indicator that economic considerations had begun to trump the notion of the enterprise as an umbrella for community life.[84] Other features of the communal infrastructure set up by the trust were at stake as well. In 1991, for instance, the directors of Shaturtorf's companies debated the future of the trust's canteens. Even though complaints about the quality of canteen food had never faded, canteens were an important social institution in workers' settlements, where they provided access to hot meals at subsidized cost. The directors clearly recognized this function. During their meeting, however, they presented canteens as an economic liability. Retaining them on the balance sheet of individual peat companies appeared increasingly impossible.[85] Irrelevant as it may seem, the discussion about Shaturtorf's canteens anticipated some of the disruption which the unmaking of the Soviet economy would entail for workers' communities. The government's haphazard move to marketization and enterprise self-financing not only disrupted the functioning, inefficient as it had been, of the Soviet economy, it also undermined the notion of the enterprise as a safety net for workers and a communal space.[86] The social contract, under which several generations of peat workers had made a home at the margins of Russia's fossil economy, was beginning to erode.

The Shatura Peat Trust had few means to withstand the economic crisis which followed the Soviet collapse and the shrinking demand for peat. Former Soviet enterprises and their employees experienced the 1990s as a period of continuous uncertainty. Left without pay for months or put on

[83] "Pribyl' vmesto ubytkov," *Za Torf* (October 5, 1990): 1.

[84] "Mysli o gazete," *Za Torf* (June 15, 1990): 1; "Prodolzhaet traditsiia," *Za Torf* (July 13, 1990): 2; "Ne rasstaius' s mnogotirazhkoi," *Za Torf* (December 21, 1990): 1.

[85] "O sanitarno-tekhnicheskom sostoianii pishchevykh ob-ektov," *Za Torf* (January 25, 1991): 1; "Stolovye ili kafe?," ibid.

[86] On the difficulties caused by the introduction of self-financing for enterprises internally and externally, see Filtzer, "Contradictions," 996–1001.

short hours, workers sought to earn money in petty trade or the informal sector. At the same time, certain of the enterprises' social responsibilities, such as medicine or pensions, were relegated to local authorities.[87] Shaturtorf's legal form changed to open joint-stock company (OAO) in 1995. While this made the company fully subject to the market, its structure and array of activities closely mirrored that of the state-owned Shatura Peat Trust in the Soviet period, including a limited commitment to provide for at least some of workers' basic needs. Eleven branches, which corresponded to the long-existing peat companies, were supposed to extract peat, produce peat-based goods, and even provide some communal services. But privatization, economic crisis, and the depletion of peat deposits in the region made it impossible for Shaturtorf to put its business on a solid footing. During the 1990s, small amounts of peat briquettes kept going to rural homes and local boiler rooms, while farming and horticulture continued using peat as a soil amendment or growing medium. Nevertheless, the demand for peat was minimal when compared to what it had been in previous decades. Unable to pay its workers, contractors, and creditors, the Shaturtorf OAO filed for bankruptcy in 2009. In 2011, it was formally liquidated.[88]

Deindustrialization affected regions marked by decades of peat extraction in uneven ways. Between 1979 and 2010, the number of people living in the city of Moscow and the Moscow oblast increased. However, areas located beyond Moscow's ever-expanding suburbs and commuter towns decreased in population, a trend visible as well in the region's most eastern districts, where the Shatura Peat Trust had been active for decades.[89] The town of Shatura had developed a diverse economy during the Soviet period, which allowed it to escape the "rustbelt realities" facing many single-enterprise towns after the Soviet collapse.[90] The Shatura Power Plant and Shatura's furniture factory, whose origins go back to

[87] Caroline Humphrey, *The Unmaking of Soviet Life: Everyday Economies after Socialism* (Ithaca: Cornell University Press, 2002), 80–82. Local managers at times tried to resist the pressures to give up factories' social welfare infrastructure, as discussed in S. V. Zhuravlev, "AvtoVAZ nachala 1990-kh gg. Istoriia privatizatsii. Kak podelit' 'Shkuru neubitogo medvedia'?," *Sotsial'naia Istoriia. Ezhegonik* (2007): 103–148, 123–125.

[88] This process is described in "Opis' no. 12 del po lichnomu sostavu Otkrytogo Aktsionernogo Obshchestva OAO Shaturtorf," Arkhivnyi otdel administratsii Gorodskogo Okruga Shatura.

[89] Grigory Ioffe and Zhanna Zayonchkovskaya, "Spatial Shifts in the Population of Moscow Region," *Eurasian Geography and Economics* 52, no. 4 (2011): 543–566, 557–559.

[90] Stephen Crowley, "Global Cities versus Rustbelt Realities: The Dilemmas of Urban Development in Russia," *Slavic Review* 79, no. 2 (2020): 365–389.

the 1960s, have continued their operations into the twenty-first century, the latter still being a major supplier for domestic customers. The Scientific Institute for Laser and Information Technologies, founded in 1979, attracted laser manufacturers to the town.[91] Outside of Shatura, however, a different trajectory unfolded. Unable to prevent social precarity and outmigration, many of the region's workers settlements, which had emerged concurrently with the peat industry, turned into what geographers have come to call Russia's "inner periphery": spaces with poor transport connectivity, little economic activity, and low population density.[92] In 2016, when I spoke to a group of women in Tugolesskii Bor, a worker's settlement just 20 kilometers to the east of Shatura, population decline was a recurrent theme during our conversation: "There's no one here anymore (*Seichas' nikogo*)," one of them said. Another commented that the local school, which used to work at capacity, stopped teaching grades 10 and upward, while the lower grades had shrunk to some 15 children per year.[93] In contrast to Tugolesskii Bor, which maintained at least some basic social infrastructure, other settlements lost almost all their inhabitants. Once controlled by the peat industry, some marginalized settlements became popular locations for Moscovites' second homes. Desertion and reappropriation sometimes occurred right next to each other. Just outside the workers' settlement of Shaturtorf, I saw derelict barracks near so-called "cottage houses" built by members of the new Russian middle class during the 2000s (Figures 6.1, 6.2, and 6.3).

Adding to the economic and social precarity that came with the peat industry's decline was a new form of ecological precarity. Abandoned peat extraction sites are highly vulnerable to wildfires, as climate change makes hot and dry summers in central Russia much more frequent.[94] Fire, of course, has accompanied the history of peat extraction since the late imperial period. Paradoxically, however, the end of extraction did not end this hazard, but amplified it even further. Milled peat extraction, the dominant form of extraction in the late Soviet period, produced vast areas

[91] *Vstrecha s Shaturoi* (Shatura: Novaia Shatura, 1996). See also, www.shatura.ru/about/story and www.shatura.com/about-company/history/. Accessed August 1, 2024.

[92] T. G. Nefedova, "Rossiiskaia periferiia kak sotsial'no-ėkonomicheskii fenomen," *Regional'nye issledovaniia* 20, no. 5 (2008): 14–30.

[93] Collective interview, Tugolesskii Bor, July 7, 2016.

[94] T. Minayeva, A. A. Sirin, and G. B. Stracher, "The Peat Fires of Russia," in Glenn B. Stracher, Anupma Prakash, and Ellina V. Sokol, eds., *Coal and Peat Fires: A Global Perspective*. Vol. 2 Photographs and Multimedia Tours (Amsterdam: Elsevier, 2013), 375–394.

FIGURE 6.1 Construction materials in front of derelict barracks outside of Shaturtorf. *Source:* Author's photograph (2019).

of deeply drained patches of land, most of which were left to themselves after exploitation ceased. Massive wildfires in 2002, but particularly in 2010, exposed the extent to which the end of extraction, population decline, and the state's gradual withdrawal from the countryside culminated in creating the conditions for a perfect storm that raged through Meshchera's scarred lands.[95]

For those who witnessed the fires on the ground, the difference in the state's approaches to wildfire in 1972 and the early twenty-first century was hard to miss. In contrast to Soviet times, when the state had mobilized workers and soldiers to fight the fires, local authorities and fire brigades received little state support in 2010.[96] Volunteers, who stepped in to fill the vacuum, could not prevent the fires from getting out of control in several parts of European Russia. The region south and east

95 T. G. Nefedova, "Khronika i prichiny goriachego leta 2010," *Znanie – Sila*, no. 6 (2011): 49–57.

96 Kontantin Novikov, "My zhivem na porokhovom sklade, so spichkami igrat' nel'zia," *Novoe Vremia* (August 16, 2010), https://newtimes.bypassnews.ru/articles/detail/25498/. Accessed August 1, 2024.

FIGURE 6.2 Newly built house outside of Shaturtorf. *Source:* Author's photograph (2019).

FIGURE 6.3 Bushes and trees encroaching on an abandoned building near Shaturtorf. *Source:* Author's photograph (2019).

of Shatura was particularly exposed to the disaster. When Oleg Yanitsky, leading environmental sociologist in Russia, interviewed forestry experts about the dramatic events, one of them responded:

> During the fire of 1972, the attitude was completely different. Firstly, people and equipment were immediately thrown here, the soldiers did not get out of here, and the forest remained intact, only the peatlands were burning. And now the peatlands, the forest, and the village have practically burned down – everything has burned down. The area is huge. This is an ecological catastrophe, a disaster. I have no idea when this will all return to normal.[97]

A major embarrassment for the Russian government, the fires of 2010 brought peatlands back onto the political agenda, at least for a short time. Unable to further ignore that central Russia's degraded peatlands represented a massive ecological liability, in August 2010, when the fires were still burning, the federal government decided to invest in nature restoration. Instead of being drained, the peatlands in some of the most badly affected parts of the Moscow region would be rewetted.[98] The decision reflected an emerging transnational consensus about the importance of peatland restoration for fire prevention, biodiversity recovery, and climate change mitigation.[99] In several locations in Russia, rewetting had been probed on a small scale already before the fateful events of 2010. In a section of the Meshchera National Park works to restore peatlands began in 2003 following widespread fires in the preceding year.[100]

Foreign support within the context of the so-called Partnership for Modernization, launched in 2010 by the European Union and Russia, turned out to be crucial to turn wetland restoration into more than a punctual one-off initiative by which the Russian government sought to publicly demonstrate its concern while the topic made headlines.[101]

97 O. N. Ianitskii, "Pozhary v 2010 g. v Rossii: Ėkosotsiologicheskii analiz," *Sotsiologicheskie issledovaniia* no. 3 (2011): 3–12, 6.

98 "Peat Bogs Flooded to Prevent Further Fires," *Moscow Times* (August 12, 2010), www.themoscowtimes.com/archive/peat-bogs-flooded-to-prevent-further-fires. Accessed July 15, 2024.

99 Joosten and Clarke, *Wise Use*, 136; J. Kreyling et al., "Rewetting Does Not Return Drained Fen Peatlands to Their Old Selves," *Nature Communications* 12 (2021): 5693, doi: https://doi.org/10.1038/s41467-021-25619-y; A. Günther et al., "Prompt Rewetting of Drained Peatlands Reduces Climate Warming Despite Methane Emissions," *Nature Communications* 11 (2020): 1644, doi: https://doi.org/10.1038/s41467-020-15499-z.

100 Akhmet'eva et al, *Torfianye bolota*, 72; Minayeva and Sirin, *Quick Scan*, 43–45.

101 "EU and Russia Launch New Partnership for Modernization," European Commission Press Release (June 1, 2010), https://ec.europa.eu/commission/presscorner/detail/en/ip_10_649. Accessed July 15, 2024.

In 2011, Germany's Chancellor Angela Merkel and Russia's President Dmitry Medvedev met in Germany to discuss bilateral collaboration in the spheres of energy, transportation, culture, and the environment. Part of a longer development that cemented Europe's fateful energy dependence on Russia and helped finance Russia's military, the meeting also led to an agreement on the restoration of peatlands. From 2011 onward, rewetting initiatives in the region of Moscow and other parts of the Russian Federation received substantial funding from the German Development Bank KfW. The scientific expertise for these projects came from the NGO Wetlands International, the Michael Succow Foundation in Greifswald, and the Russian Academy of Sciences. As a result, more than 73,000 hectares have been rewetted in the region of Moscow, a scale of peatland restoration with few parallels globally.[102]

Burning swamps haunt the country despite these achievements. Rewetting initiatives in the wider Moscow area have certainly reduced the local wildfire risk and carbon dioxide emissions, but degraded peatlands continue to pose a hazard in other parts of Russia. Most of them are localized events, but it happens regularly that such fires interfere with people's everyday lives or that they persist for several months. In October 2021, smoke from burning peatlands forced the authorities in the city of Ekaterinburg in the Urals to close schools and divert airplanes. Some fires in the region smoldered throughout the winter and into the following spring. Only in April 2022 did local media report that they had been finally extinguished.[103] Fires make degraded peatlands both sites and agents of the climate crisis. And they expose the otherwise invisible ways in which the history of peat extraction in central Russia reverberates in places far away. Even undrained or unexploited peatlands are now becoming increasingly unstable – particularly in the Russian Arctic, where the changing climate unsettles scientific and popular understandings of permafrost and where discontinuity is a material process as much as a

[102] A. A. Sirin et al., "Multispectral Satellite based Monitoring of Land Cover Change and Associated Fire Reduction after Large-scale Peatland Rewetting Following the 2010 Peat Fires in Moscow Region (Russia)," *Ecological Engineering* 158 (2020): 106044, 2; www.succow-stiftung.de/en/peatlands-climate/moorrestauration-in-russland. Accessed July 15, 2024.

[103] "Burning Peat and Wildfires Bring Unseasonal Smoke Apocalypse to Urals and Siberia," *The Siberian Times* (October 14, 2021), https://siberiantimes.com/other/others/news/burning-peat-and-wildfires-bring-unseasonal-smoke-apocalypse-to-urals-and-siberia/; "Byl primenen novyi metod: v Sverdlovskoi oblasti za zimu okonchatel'no potushili vse tleiushchie torfianiki," *E1-Ru* (April 16, 2022), www.e1.ru/text/incidents/2022/04/16/71262008/. Accessed July 15, 2024.

social experience.[104] Thawing peatlands emit substantial quantities of the highly potent GHG methane. They also become more vulnerable to fire, which makes them potential sources of considerable future carbon dioxide emissions.[105] As the earth is entering uncharted territory, remembering the forgotten margins of Russia's fossil economy is one way of noticing the varied histories and countless expressions of the planetary emergency.

[104] Chu, *The Life of Permafrost*, 164–175. On discontinuity as a process and a lens of inquiry, see Wrigley, *Earth, Ice, Bone, Blood*, 30–33.

[105] Turetsky et al., "Global Vulnerability"; Alexandra Witze, "The Arctic Is Burning Like Never Before: And That's Bad News for Cimate Change," *Nature* 585 (2020): 336–337.

Conclusions

Burning Swamps has situated the history of peat in Russia's fossil economy in specific locations and moments of time. But this history has no clear boundaries. Like many other extractive industries, peat extraction left many regions of Russia with an "unruly heritage" that forms an active presence long after extraction ended.[1] In the spring of 2023, Andrei Kuriakov, then Vice-Minister for Natural Resources and the Environment of the Sverdlovsk oblast in the Urals, outlined his strategies to prevent the recurring peat fires which had regularly beset the area in preceding years. Kuriakov identified three possible ways to address the problem: rewetting the peatlands, exploiting them until the complete exhaustion of the peat layer, or monitoring them with cameras to allow firefighters to respond to any sign of fire without delay. Informed by a narrow concern for fire prevention, the proposal omitted the possibilities of wetland restoration for the preservation of biodiversity or climate change mitigation. In fact, the idea of continuing peat extraction suggests an appreciation for peatlands as resources rather than as ecosystems. Nonetheless, there is something noteworthy in the official's framing of the problem. To determine fire-prone sites, Kuriakov advised that regional authorities should consult Soviet archives to identify previously

[1] On the concept of "unruly heritage" and its significance for studying the physical remains of socialist extractivism, see Anatolijs Venovcevs, "Living with Socialism: Toward an Archaeology of a Post-Soviet Industrial Town," *The Extractive Industries and Society* 8, no. 4 (2021): 100835.

exploited peat deposits.[2] As fires haunt Russia's degraded peatlands, remembering their history becomes imperative to preventing disaster.

The official's desperate turn to the archive points to the fragmented memory of peat extraction in recent decades. Typically, this history only receives attention when fire and smoke make it tangible for people far away. Memory operates differently in areas immediately affected by past extraction though. In Shatura, peat has long been part of the official historical narrative. When I went there in 2016, the Museum of Local Lore invited visitors to read the history of Shatura and the Meshchera through the lens of peatlands and peat workers. But while the exhibition drew a clear boundary between past and present, the geographies of settlement, transport, and energy in and around Shatura tell a different story. Like most remnants of discontinued extractive and industrial activities in Russia, the sites and things created by the peat industry are not formally recognized as industrial heritage.[3] In parts of the Shatura region, trees, shrubs, and grasses encroach on former train depots, as workers' former houses are left to decay. Where exploited peatlands have been rewetted, reading the landscape as a result of extractive labor becomes ever more challenging for the untrained eye (Figure 7.1). Abandonment and restoration do not mark the end of history, however, but rather incorporate elements of it in new ways into the present. Most narrow-gauge railway tracks have been disassembled, but their paths still shape people's mobilities as former tracks have become country roads. Fishermen and campers flock to the shores of rewetted peat extraction sites. Derelict buildings and infrastructure attract informal heritage tourists and the local youth.[4] Long after it has ceased, peat extraction keeps shaping how people interact with the local environment.

Reminders of the lingering presence of the past, the places created by peat extraction are emblematic of the myriad pathways that led Russia,

[2] "Gorit v vode i pod zemlei: Naskol'ko velika ugroza torfianykh pozharov i chto s nimi delat'?" *Moskovskaia Gazeta* (April 25, 2023), https://mskgazeta.ru/obshchestvo/gorit-v-vode-i-pod-zemlej-naskol-ko-velika-ugroza-torfyanyh-pozharov-i-chto-s-nimi-delat–11914.html. Accessed August 1, 2024.

[3] N. A. Rodionov, "Torfopromyshlennoe nasledie Urala XX veka: Problemy izucheniia i sokhraneniia," *Bakuninskie chteniia: Sektsiia 5. Istoriia nauki i tekhniki* (2019): 344–355.

[4] Elena Kochetkova and Anna Petrova, "Abandoned, But Not Forgotten Heritage: Former Industrial Enterprises in Cultural and Urban Russian Landscapes," *Heritage & Society* 15, no. 3 (2022): 277–298. For one example of such heritage tourism in the Shatura region, see "Istoriia poslednei uzkokoleinoi dorogi Pod'moskov'ia – Shaturskoi UZhD," https://pikabu.ru/story/istoriya_posledney_uzkokoleynoy_dorogi_podmoskovya__shaturskoy_uzhd_8002949?utm_source=linkshare&utm_medium=sharing. Accessed July 19, 2024.

FIGURE 7.1 Rewetted peat extraction site near Shatura. *Source*: Author's photograph (2019).

one of the world's major carbon debtors, into the age of fossil fuels. Energy historians have had good reasons to focus their narratives on coal, oil, and gas, but it is crucial to remember the importance of less iconic and less efficient fuels. From the late nineteenth century, central Russia's wetlands became industrial hinterlands serving the energy demands of some of the empire's largest enterprises. They retained this role into the latter part of the twentieth century, as electric power plants, mostly built in the early Soviet period, consumed massive amounts of peat. The labor of thousands of women and men proved crucial in mobilizing these carbon stocks. Traveling back and forth between their villages and the extraction sites, peat workers became part of Russia's expanding industrial metabolism while retaining close ties with the village economy. From the 1950s, machines replaced seasonal workers, and permanent workers turned the margins of Russia's fossil economy into their homes. Still, even for that later period, peat extraction sits awkwardly within established frameworks of social and economic history. It remained subject to the seasons and the weather, while peat workers spent much of their time producing food on private plots and auxiliary farms. Neat distinctions between agriculture and industry, organic and fossil energy regimes, production and reproduction, as well as capitalism and state socialism become blurred once we focus on these forgotten types of matter and labor in Russia's burgeoning fossil economy.

But this history is part of yet another story. Scientists have pointed to the massive destruction of wetland ecosystems globally, a process that gained traction since the early eighteenth century and accelerated significantly during the twentieth. Many plants and animals lost their habitat, and vast amounts of soil carbon have been released into the atmosphere. From a birds-eye perspective, wetlands in Russia have escaped the modern onslaught on these ecosystems. Less than 10 percent of the wetlands within the country's present borders have been destroyed; a stark contrast to Ireland, for instance, which has lost more than 90 percent of its wetland coverage. However, the picture changes when focusing on the absolute area of destruction instead of expressing it as a percentage of the overall national territory. Alongside the United States, China, India, and Indonesia, Russia ranks among the five countries responsible for 40 percent of global wetland loss.[5] Within the country, the geography of wetland destruction is highly uneven. While large parts of Siberia's peatlands have escaped drainage and extraction, up to half of the peatlands in parts of central, western and north-western Russia are degraded.[6] Ironically, it is by taking a regional perspective that the planetary significance of peat in Russia's fossil economy becomes evident.

The future of peatlands in Russia is uncertain. As I am completing this book, the country is invested in a brutal war of aggression against neighboring Ukraine, deliberately targeting its civilian population, energy infrastructure, and cultural heritage as well as environments that are vital for both human and nonhuman life in Ukraine.[7] As Russia's resources are wasted in warfare, it is unlikely that repairing or protecting the country's peatlands will receive much attention in the coming years. At the same time, rewetting projects have lost the international funding that had been key to earlier advancements.[8] While peatland restoration sometimes

[5] Etienne Fluet-Chouinard et al, "Extensive Global Wetland Loss over the Past Three Centuries," *Nature* 614 (2023): 281–286.

[6] Sirin et al., "Multispectral Satellite based Monitoring," 1.

[7] For reflections on how to frame the war through interspecies relationships, see Tanya Richardson, "Interspecies Relations in the Midst of the Russia–Ukraine War," *Current History* 122, no. 847 (2023): 301–307.

[8] www.succow-stiftung.de/en/peatlands-climate/moorrestauration-in-russland. Accessed August 1, 2024.

continues with local means, several companies in Russia are invested in peat extraction, in part to meet China's growing demand for agricultural peat.[9] Since the beginning of Russia's full-scale invasion of Ukraine in 2022, environmental regulations, which had been weak and insufficient, have been further undermined, while the state's crackdown on independent media, intimidation, and repression make public resistance against pollution, extractivism, or waste dumping increasingly difficult and dangerous.[10] Greenpeace was declared an "undesirable organization" in May 2023, forcing the NGO to end its presence in the country.[11] A significant blow to environmental activism in Russia, this development may also directly affect the ability to prevent and combat wildfires. Greenpeace Russia had developed unique expertise in monitoring and extinguishing peat fires, often filling the gap left by insufficient state support for fire brigades.[12] The organization also made itself a name calling out the failure of regional administrations to respond to peat fires.[13] As climate change makes Eurasian summers hotter and dryer, the void left by the crackdown on Greenpeace could translate into more intense fire events on Russia's degraded peatlands.

[9] "Rossiiskie uchenye nashli reshenie problemy zabroshennykh torfianikov," *RIA Novosti* (November 9, 2023), https://ria.ru/20231109/nauka-1908171629.html?utm_source=yxnews&utm_medium=desktop; "Na puti k spaseniiu redkikh vidov: Kak v Rossii vosstanavlivaiut torfianye bolota," *Federal Press Leningradskaia Oblast'* (October 8, 2023), https://fedpress.ru/article/3272460; "Ural'skaia kompaniia nachnet postavku torfa v Kitai," *TASS* (June 15, 2019), https://tass.ru/ekonomika/6554125; "Viatka Torf provel peregovory s kitaiskim biznesom po postavkam sel'khoztorfa v KNR," *Kirovskaia Pravda* (August 30, 2023), https://kirovpravda.ru/vyatkatorf-provel-peregovory-s-kitayskim-biznesom-po-postavkam-selhoztorfa-v-knr/?ysclid=lpu91ey29v812680409. Accessed August 1, 2024.

[10] Zinaida Palvinskaya, "How War Affects Grassroots Environmental Protest," *Environment & Rights* 85 (October 2022): 36–41; Angelina Davydova, "Russische Klimaaktivist:innen: Geächtet, bedroht, aber noch da," *taz* (September 16, 2023), https://taz.de/Russische-Klimaaktivistinnen/!5956935/. Accesses March 27, 2025.

[11] "Environmental Protection Is 'Undesirable' in Russia," *Greenpeace International* (May 19, 2023), www.greenpeace.org/international/press-release/59745/environmental-protection-undesirable-russia-greenpeace/. Accessed July 15, 2024.

[12] The organization's guidelines for dealing with peat fires, first published in Russian in 2020, were translated into English in 2022. See *Recommendations on How to Extinguish Peat Fires in Drained Peatlands: The Russian Experience. Edition adapted for global perspective* (Greenpeace, 2022), doi: 10.35075/GPWFIRE.2022.79.74.001.

[13] For one example, see "Grinpis: Torfianye pozhary v Branskoi oblasti prodolzhaiutsia, no ob ikh masshtabakh molchat," *Nash Briansk* (2014), https://news.nashbryansk.ru/2014/10/23/hot-hour/grinpis-torfyanye-pozhary-v-bryanskoj-oblasti-prodolzhayutsya-no-ob-ix-masshtabax-molchat-88413/. Accessed July 15, 2024.

Remembering the forgotten margins of Russia's fossil economy allows us to see the destruction of, as well as the promise lying in, peatlands elsewhere. Cultural perceptions of wetlands have shifted tremendously in recent decades. After being regarded as wastelands for centuries, they are now valued for their role as habitats, flood regulators, and carbon stores and sinks.[14] Peatlands, the most widespread type of wetland in the world, have become a subject of popular nature-writing as scientists address a broader public to communicate the importance of protecting or rehabilitating these landscapes.[15] Russia may not soon become an ally in efforts to cure degraded peatlands from the damage inflicted on them over the past centuries. But the world is incredibly rich in destroyed peatlands. Many are degraded so severely that they are hardly recognizable as peatlands anymore. Writing their varied histories is the first step in addressing the ecological amnesia surrounding peatland ecosystems and noticing the possibility for new forms of more-than-human co-existence amidst the fragility of the current moment. As the anthropologist Anna Tsing wrote, "in a global state of precarity, we don't have choices other than looking for life in this ruin."[16] The exploitation of peatlands has undoubtedly made life on earth more vulnerable. But there is loss, and there is chance in this history.

[14] As the environmental historian John R. McNeill put it, wetland restoration is "perhaps the best example of how cultural preferences with respect to a given biome can change over time, and the reviled can become the revered." John R. McNeill, "The State of the Field of Environmental History," *Annual Review of Environment and Resources* 35 (2010): 345–374, 364.

[15] Franziska Tanneberger and Vera Schroeder, *Das Moor: Über eine faszinierende Welt zwischen Wasser und Land und warum sie für unser Klima so wichtig ist* (Munich: dtv, 2023); Annie Proulx, *Fen, Bog & Swamp: A Short History of Peatland Destruction and Its Role in the Climate Crisis* (London: 4th Estate, 2022); *Peatland Atlas: Facts and Figures about Wet Climate Guardians* (Berlin: Heinrich-Böll-Stiftung et al., 2023).

[16] Tsing, *Mushroom at the End of the World*, 6.

Bibliography

ARCHIVAL COLLECTIONS CITED

Russian State Historical Archive, Saint Petersburg (RGIA)

f. 424 Zapadnaia Ėkspeditsiia po osusheniiu bolot
f. 425 Glavnoe Upravlenie Zemleustroistva i Zemledeliia – Torfmeisterskaia chast'

Russian State Archive of the Economy, Moscow (RGAĖ)

f. 99 Nauchno-issledovatel'skii ėkonomicheskii institut Gosplana Soveta Ministrov SSSR
f. 356 I. I. Vikhliaev papers
f. 399 Sovet po izucheniiu proizvoditel'nykh sil (SOPS) pri Gosplane SSSR
f. 478 Narodnyi komissariat zemledeliia RSFSR (Narkomzem RSFSR)
f. 544 Komissiia po zapovednikam i okhrane prirody i laboratoriia okhrany prirody
f. 4372 Gosudarstvennyi planovoi komitet SSSR (Gosplan SSSR)
f. 5208 Gosudarstvennaia komissiia po ėlektrifikatsii Rossii (GOĖLRO)
f. 7486 Ministerstvo sel'skogo khoziaistva SSSR
f. 7969 Glavnoe upravlenie torfianoi promyshlennosti Ministerstva ėlektrostantsii SSSR
f. 9455 I. I. Radchenko papers
f. 9508 R. Ė. Klasson papers
f. 9458 P. V. Taneev papers

State Archive of the Russian Federation, Moscow (GARF)

f. A-259 Sovet Ministrov RSFSR
f. A-262 Gosudarstvennyi Planovoi Komitet RSFSR (Gosplan RSFSR)

f. A-314 Glavnoe upravlenie kommunal'nogo khoziaistva pri Soviet Narodnykh Komissarov RSFSR
f. A-385 Verkhovnyi Sovet RSFSR
f. A-412 Glavnoe upravlenie torfianogo fonda (Glavtorffond)
f. A-632 Ministerstvo toplivnoi promyshlennosti RSFSR
f. R-5446 Sovet Ministrov SSSR
f. R-7952 Gosudarstvennoe izdatel'stvo "Istoriia fabrik i zavodov"

Archive of the Russian Academy of Sciences, Moscow (ARAN)

f. 1920 N. I. P'iavchenko papers
f. 2067 Nauchnyi Sovet po probleme biogeotsenologii i okhrany prirody

Online Repositories

The Harvard Project on the Soviet Social System Online (https://library.harvard.edu/sites/default/files/static/collections/hpsss/index.html)
Prozhito: Lichnye istorii v ėlektronnom korpuse dnevnikov i vospominanii (https://corpus.prozhito.org/)

PUBLISHED PRIMARY SOURCES

(a) Journals and Newspapers

Lesnoi zhurnal
Pravda
Torfianaia promyshlennost'
Torfianoe delo
Trudy Vol'nogo ėkonomicheskogo obshchestva
Vestnik torfianogo dela
Za Torf
Zhurnal Ministerstva gosudarstvennykh imushchestv

(b) Books, Articles, Collections

VIII International Peat Congress: Proceedings. Vol. II (Leningrad, 1988).
XVIII S-ezd Vsesoiuznoi Kommunisticheskoi Partii (b) 10–21 marta 1939 g. Stenografichekii otchet (Moscow: OGIZ, 1939).
Abramova, T. G., M. S. Boch and E. A. Galkina, eds.: *Tipy bolot SSSR i printsipy ikh klassifikatsii* (Leningrad: Nauka, 1974).
Artel'nyi dogovor ob usloviiakh oplaty i poriadke rabot torfianokov v sezon 1925 goda (Moscow: VSNKh – Gostorf 1925).
Atlas Moskovskoi Oblasti (Moscow: Glavnoe Upravlenie Geodezii i Kartografii, 1964).

Atlas Moskovskoi Oblasti (Moscow: Glavnoe Upravlenie Geodezii i Kartografii, 1976).

Boch, M. S.: "Vsesoiuznoe soveshchanie po voprosam okhrany bolot (Leningrad, 15 maia 1974g.)," *Botanicheskii zhurnal* 61, no. 2 (1976): 291–295.

Boch, M. S. and V. V. Mazing: *Ėkosistemy bolot SSSR* (Leningrad: Nauka, 1979).

Bukhgeim, E. O.: *K ėkonomicheskomu osvobozhdeniiu Rossii putem ėlektrifikatsii ee territorii* (Moscow: S. P. Iakovlev, 1915).

Chekhov, Anton: *Plays*. Translated with Notes by P. Carson. With an Introduction by R. Gilman (London: Penguin Books, 2002).

Dlugotskii, L. I., P. V. Dolinin and N. N. Uspenskii: *Frezernyj torf (Po materialam komissii SOIUZTORF po frezernomu torfu)* (Moscow: Nauchno-issled. torf. institut, 1931).

Dobreitser, I.: "Bor'ba s maliariei na torforazrabotkakh," *Izvestiia Narodnogo Komissariata Zdravokhraneniia* 6, no. 4–5 (1923): 13–14.

Dorofeev, Vil', Valerii Povoliaev and Vladimir Travinskii: "Skhvatka s ognem," *Literaturnaia Gazeta* no. 33 (16 August 1972): 13.

Dovlatov, Sergei: *The Zone*. Translated by Anne Frydman. Edited by Katherine Dovlatov (London: OneWorld Classics, 2011).

Ėlektrifikatsiia Tsentral'nogo Raiona: Sostavleno Gosudarstvennoi Kommissiei po Ėlektrifikatsii Rossii (Moscow: Gosudarstvennoe Tekhnicheskoe Izdatel'stvo, 1920).

Ėntsiklopedicheskii slovar', tom IV (Saint Petersburg: Brokgauz and Efron, 1891).

Fabrichno-zavodskaia promyshlennost' i torgovlia Rossii (Saint Petersburg: Ministerstvo Finansov, 1896).

Gorshkov, L. A., et al.: *Kompleksnoe ispol'zovanie torfianykh bolot* (Moscow: Rossel'khozizdat, 1969).

Goskomstat, *Naselenie SSSR 1988: Statisticheskii ezhegodnik* (Moscow: Finansy i statistika, 1989).

GUZiZ: *Ezhegodnik Otdela zemel'nykh uluchshenii, 1912* (Saint Petersburg: Tipografiia V. F. Kirshbauma, 1913).

Ezhegodnik Otdela zemel'nykh Uluchshenii, 1913 (Petrograd: Tipografiia V. F. Kirshbauma, 1914).

Obzor deiatel'nosti za 1913 god (Petrograd: Tipografiia Sel'skogo Vestnika, 1915).

Otchet o rabotakh Torfmeisterskoi chasti v 1913 god (Petrograd: Tipografiia Al'tshulera, 1915).

Illiustrirovannoe opisanie Vserossiiskoi khudozhestvenno-promyshlennoi vystavki v Moskve 1882 g. (Saint Petersburg: Tipografiia Ėduarda Goppe, 1882).

Ilovaiskii, S. I.: *Uchebnik finansovogo prava: 4-e izdanie* (Odessa: Tipo-Khromo-Lit. A. F. Sokolovskago, 1904).

"International Conference on the Conservation of Wetlands and Waterfowl: Convention on Wetlands of International Importance," *International Legal Materials* 11, no. 5 (1972): 963–976.

Iskrin, D. I.: *Na torfianykh bolotakh: Bytovoi ocherk po pesniam, zapisannym v Bogorodskom uezde* (Bogorodsk: Bogorodskii Institut Kraevedeniia i UProsa, 1925).

Ivanov, K. E.: *Vodoobmen v bolotnykh landshaftakh* (Leningrad: Gidroprometeoizdat, 1975).
Kafengauz, L. B.: *Ėvoliutsiia promyshlennogo proizvodstva Rossii (posledniaia tret' XIX v. –30-e gody XX v.)* (Moscow: Ėpifaniia, 1994).
Karsakov, V.: "Chem otaplivat' parovozy? Zametka o vozmozhnom razvitii torfianogo dela v sviazi s nuzhdami gosudarstvennogo i zheleznodorozhnogo khoziaistva," in P. M. Solov'ev, ed., *Torfianoe delo: Obzor vydaiushchikhsia novostei po kul'ture bolot i torfodobyvaniiu za poslednee vremia* (Saint Petersburg: Tipografiia Berman i Ko, 1897), 68–79.
Khlevniuk, O. V., V. A. Kozlov and S. V. Mironenko, eds., *Zakliuchennye na stroikakh kommunizma: Gulag i ob-ekty ėnergetiki v SSSR. Sobranie dokumentov i fotografii* (Moscow: ROSSPĖN, 2008).
Klasson, R. Ė.: "Gidrotorf v sviazi s raionnymi ėlektricheskimi stantsiiami," *Ėlektrichestvo* no. 5–6 (1923): 257–266.
"Hydrotorf," *Zeitschrift des Vereines deutscher Ingenieure* 68, no. 23 (7 June 1924): 601–605.
Kol'tsov, N. K.: *Bolotnaia likhoradka (Maliariia) i komary* (Moscow: Tipogr. Lomonosov, 1913).
Kratkii ocherk rabot ėkspeditsii po osusheniiu bolot v Tsentral'nom raione (Saint Petersburg: Zapadnaia Ėkspeditsiia po osusheniiu bolot, 1896).
Lenin, V. I.: *Collected Works Vol. 3: The Development of Capitalism in Russia* (Moscow: Progress Publishers, 4th printing 1977).
Collected Works Vol. 31: April–December 1920 (Moscow: Progress Publishers, 2nd printing 1974).
Levchenko, Ia. P.: *Planirovka rabochikh poselkov* (Moscow: Izdatel'stvo NarkomKhoza RSFSR, 1940).
Lipenskii, G.: *Bol'shaia Shatura* (Moscow: Moskovskii Rabochii, 1970).
Lomonosov, M.: *O sloiakh zemnykh i drugie raboty po geologii: S predisloviem prof. G. G. Lemmleina* (Moscow: Gosudarstvennoe izdatel'stvo geologicheskoi literatury, 1949).
Martsinovskii, E. I. and V. I. Martsinovskii, "Torfodobyvaiushchaia promyshlennost' i maliariia," *Gigiena i Ėpidemiologiia* 2, no. 1 (1923): 60–67.
Martsinovskii, V. I.: *Maliariia v torfodobyvaiushchei promyshlennosti i bor'ba s nei* (Moscow: TsUTorf, 1923).
Maslov, B.: "Esli v kolodtse ischezla voda ...," *Literaturnaia Gazeta*, no. 7 (1983): 13.
Maslov, B. S. and I. V. Minaev: *Melioratsiia i okhrana prirody* (Moscow: Rossel'khozizdat, 1985).
Nabokov, Vladimir: *Mary*. Translated from the Russian by Michael Glenny, in collaboration with the author (New York: Vintage International, 1989).
Nikitinskii, N. A.: *Torf i ego razrabotka na toplivo* (Moscow: Izdanie politekhnicheskogo obshchestva, 1882).
Nitsenko, A. A. and M. S. Boch, eds., *Priroda bolot i metody ikh issledovanii* (Leningrad: Nauka, 1967).
Oganov, L. I.: "Maliariia v Moskovskoi oblasti i ee sezonnaia periodichnost'" (Dissertatsiia na soiskanie uchenoi stepeni kandidata meditsinskikh nauk,

Moskovskii oblastnoi imeni Mechnikova institut ėpidemiologii, mikrobiologii i infektsionnykh boleznei, 1947).

Okuneva, I. and E. Shteinbakh, *Ratsionalizatsiia i ozdorovlenie truda torfianits* (Moscow: Gosudarstvennoe sotsial'no-ėkonomicheskoe izdatel'stvo, 1931).

Otchet Glavnogo Torfianogo Komiteta: Dobycha torfa v 1919 g. i snabzhenie prodovol'stviem torfianikov za tot zhe period (Moscow: VSNKh, 1920).

Plan ėlektrifikatsii RSFSR: Vvedenie k dokladu 8-mu S-ezdu Sovetov Gosudarstvennoi Komissii po Ėlektrifikatsii Rossii (Moskva: Gosudarstvennoe tekhnicheskoe izdatel'stvo 1920).

Prishvin, M. M.: *Sobranie sochenenii, Vol. 3, Proizvedeniia 1924–1935 godov* (Moscow: Khudozhestvennaia Literatura, 1983).

Proizdvoditel'nye sily Rossii: Kratkaia kharakteristika razlichnykh otraslei truda – sootvetstvenno klassifikatsii vystavki (Saint Petersburg: Ministerstvo Finansov, 1896).

P'iavchenko, N. I., ed., *Vzaimootnosheniia lesa i bolota: Po dannym statsionarnykh issledovanii* (Moscow: Nauka, 1967).

Ramzin, Leonid K.: "The Power Resources of Russia," in *The Transactions of the First World Power Conference 1924: Vol. 1 Power Resources of the World Available and Utilised* (London: Percy Lund Humphries & Co, 1925).

Raschetin, G. A.: *Profilaktika i tushenie torfianykh pozharov* (Moscow: Stroiizdat, 1964).

"Russkie bolota," *Vestnik Evropy* 8, no. 12 (1873): 754–773.

Sbornik statisticheskikh svedenii po Moskovskoi gubernii: Vypusk II: Raspronstranenie i proiavlenie maliiarii v naselenii Moskovskoi gubernii, ed. A. I. Skibnevskii (Moscow: Pechatnia S. P. Iakovleva, 1903).

Sbornik svedenii po departamentu zemledeliia i sel'skoi promyshlennosti (Saint Petersburg: Ministerstvo Gosudarstvennykh Imushchestv, 1879).

Sdelaem Rossiiu ėlektricheskoi: Sbornik vospominanii uchastnikov Komissii GOĖLRO i stroitelei pervykh ėlektrostantsii (Moscow: Gosudarstvennoe ėnergeticheskoe izdatel'stvo, 1961).

Semenova-Tian-Shanskaia, A. M. and N. A. Frantskevich, eds., *Voprosy okhrany botanicheskikh ob-ektov*, (Leningrad: Nauka, 1971).

Shablinskii, V. V.: *Osushenie torfianykh mestorozhdenii* (Moscow: Gosudarstvennoe ėnergeticheskoe izdatel'stvo, 1953).

Shatura: Sbornik statei k 25-letiiu GRĖS imeni V. I. Lenina (Moscow, 1951).

Shaturskaia GRĖS (Moscow: Informėnergo, 1970).

Simmons, Cynthia and Nina Perlina, eds., *Writing the Siege of Leningrad: Women's Diaries, Memoirs and Documentary Prose* (Pittsburgh: University of Pittsburgh Press, 2002).

Sokolov, V. N.: *Torf v narodnom khoziaistve* (Moscow: Nedra, 1988).

Solzhenitsyn, Alexander: *Matryona's House and Other Stories*. Translated by Michael Glenny (Harmondsworth: Penguin Books, 1974).

One Day in the Life of Ivan Denisovich. Translated from the Russian by H. T. Willetts. Foreword by Alexis Klimoff (London: Vintage Books, 2003).

"Sovershenno sekretno": Liubianka–Stalinu o polozhenii v strane (1922–1934 gg.). 10 volumes (Moscow: IRI RAN, 2001–2017).

Sukachev, V. N.: *Bolota: Ikh obrazovanie, razvitie i svoistva*, 2nd ed. (Petrograd: Novaia Derevnia, 1923).

Time of Troubles: The Diary of Iurii Vladimirovich Got'e, Moscow, July 8, 1917 to July 23, 1922. Translated, edited and introduced by Terence Emmons (Princeton: Princeton University Press, 1988).

Torfianoe delo v RSFSR i deiatel'nost' Torfianoi sektsii za period 1923–1924 gg. (Moscow: Narkomzem, 1925).

Torfianoi fond RSFSR: Postanovleniia, polozheniia i itogi rabot po gosudarstvennomu torfianomu fondu (Moscow: Narkomzem, 1945).

Torfianoi fond RSFSR: Torfianye bolota Moskovskoi oblasti (Moscow: Narkomzem, 1940).

Troepol'skii, G.: "O rekakh, pochvakh i prochem," *Novyi Mir* 41, no. 1 (1965): 185–212.

Trotsky, Leon: *Problems of Everyday Life and Other Writings on Culture & Science* (New York: Pathfinder Press, 4th printing 1986).

Trudy 1-ogo s-ezda vrachei torforazrabotok 8-10 Ianvaria 1926 (Skopin: Narkomzdrav-Gostorf, 1926).

Trudy Kommissii po izucheniiu torfianogo dela (Moscow: Obshchestvo dlia sodeistviia uluchsheniiu i razvitiiu manufakturnoi promyshlennosti, 1916).

Trudy soveshchanii po podmoskovnomu ugliu i torfu, sozvannykh v Moskve na 20–22 noiabria 1915 goda Moskovskim upolnomochennym predsedatelia Osobogo soveshchaniia po toplivu (Moscow: Biuro ob-edinennykh tekhn. organizatsii, 1916).

Trudy vtorogo soveshchaniia po voprosam bor'by s maliariei sredi torforabochikh, 11–13 fevralia 1928 (Moscow: Gosudarstvennoe meditsinskoe izdatel'stvo, 1928).

Vronskii, Iu.: "Ostorozhno boloto," *Literaturnaia Gazeta*, no. 14 (1976): 2.

Zaikov, A. M.: *Geroicheskaia Shatura: Pervenets Sovetskoi ėlektrifikatssii* (Moscow: Gosudarstvennoe izdatel'stvo, 1928).

Zapadnaia ėkspeditsiia po osusheniiu bolot, *Kratkii ocherk rabot ėkspeditsii po osusheniiu bolot v tsentral'nom raione* (Saint Petersburg: Tipografiia V. Bezobrazova, 1896).

Zverkov, Iu. V.: *Vtoraia zhizn' torfianikov* (Kirov: Volgo-Viatskoe kn. izd-vo, 1982).

SECONDARY LITERATURE

Akhmet'eva, N. P., et al.: *Torfianye bolota tsentral'nykh oblastei Evropeiskoi chasti Rossii: Ikh transformatsiia za poslednie 100 let* (Moscow: GEOS, 2020).

Albritton Jonsson, Fredrik and Carl Wennerlind: *Scarcity: A History from the Origins of Capitalism to the Climate Crisis* (Cambridge, MA: Harvard University Press, 2023).

Aleshinskaia, A. S., et al.: *Drevnie okhotniki i rybolovy Podmoskov'ia po materialam mnogosloinogo poseleniia ėpokhi kamnia i bronzy – Voimezhnoe I* (Moscow: RAN Institut Arkheologii, 1997).

Allen, Robert C.: "The Rise and Decline of the Soviet Economy," *Canadian Journal of Economics* 34, no. 4 (2001): 859–881.

Andrews, Thomas G.: *Killing for Coal: America's Deadliest Labor War* (Cambridge, MA: Harvard University Press, 2008).

Arndt, Melanie and Laurent Coumel: "A Green End to the Red Empire? Ecological Mobilizations in the Soviet Union and Its Successor States, 1950–2000: A Decentralized Approach," *Ab Imperio*, no. 1 (2019): 105–124.

Asschenfeldt, Friedrich and Max Trecker: "From Ludendorff to Lenin? World War I and the Origins of Soviet Economic Planning," *Europe-Asia Studies* 76, no. 1 (2024): 10–28.

Ashwin, Sarah: "Endless Patience: Explaining Soviet and Post-Soviet Social Stability," *Communist and Post-Communist Studies* 31, no. 2 (1998): 187–198.

Bailes, Kendall E.: *Technology and Society under Lenin and Stalin: Origins of the Soviet Technical Intelligentsia, 1917–1941* (Princeton: Princeton University Press, 1978).

Balmaceda, Margarita M.: *Russian Energy Chains: The Remaking of Technopolitics from Siberia to Ukraine to the European Union* (New York: Columbia University Press, 2021).

Banerjee, Anindita: "Electricity: Science Fiction and Modernity in Early Twentieth-Century Russia," *Science Fiction Studies* 30, no. 1 (2003): 49–71.

Barak, On: *Powering Empire: How Coal Made the Middle East and Sparked Global Carbonization* (Oakland: University of California Press, 2020).

Barca, Stefania: *Forces of Reproduction: Notes for a Counter-Hegemonic Anthropocene* (Cambridge: Cambridge University Press, 2020).

"Labouring the Earth: Transnational Reflections on the Environmental History of Work," *Environment and History* 19, no. 1 (2014): 3–27.

Barenberg, Alan: *Gulag Town, Company Town: Forced Labor and Its Legacy in Vorkuta* (New Haven: Yale University Press, 2014.

Baron, Nick: "New Spatial Histories of Twentieth Century Russia and the Soviet Union: Surveying the Landscape," *Jahrbücher für Geschichte Osteuropas* 55, no. 3 (2007): 374–400.

Barragán Romano, Rossana and Leda Papastefanaki: "Women and Gender in the Mines: Challenging Masculinity through History: An Introduction," *International Review of Social History* 2, no. 65 (2020): 191–230.

Bassin, Mark: "Geographical Determinism in Fin-de-siècle Marxism: Georgii Plekhanov and the Environmental Basis of Russian History," *Annals of the Association of American Geographers* 82, no. 1 (1992): 3–22.

Bassin, Mark, Christopher Ely, and Melissa K. Stockdale, eds., *Space, Place, and Power in Modern Russia: Essays in the New Spatial History* (DeKalb: Northern Illinois University Press, 2010).

Beckert, Sven, Bosma Ulbe, Mindi Schneider, and Eric Vanhaute: "Commodity Frontiers and the Transformation of the Global Countryside: A Research Agenda," *Journal of Global History* 16, no. 3 (2021): 435–450.

Bellamy, David J. and Tom Pritchard: "Project 'Telma': A Scientific Framework for Conserving the World's Peatlands," *Biological Conservation* 5, no. 1 (1973): 33–40.

Bernstein, Frances L.: "'The Dictatorship of Sex': Science, Glands, and the Medical Construction of Gender Difference in Revolutionary Russia," in David L. Hoffmann and Yanni Kotsonis, eds., *Russian Modernity: Politics, Knowledge, Practices* (New York: St Martin's Press, 2000), 138–160.

Beuerle, Benjamin: "A Dormant Giant: Renewable Energy in the Soviet Union and Russia (1970s–Present)," *Russian Review* 84, no. 2 (2025): 267–282.

Bird, Robert: "The Poetics of Peat in Soviet Literary and Visual Culture, 1918–1959," *Slavic Review* 70, no. 3 (Fall 2011): 591–614.

Blackbourn, David: *The Conquest of Nature: Water, Landscape and the Making of Modern Germany* (London: W. W. Norton, 2006).

Blokov, Ivan: "Alle Jahre wieder: Waldbrände in Russland," *Osteuropa* 60, no. 9 (2010): 5–16.

Bolotova, Alla: "Loving and Conquering Nature: Shifting Perceptions of the Environment in the Industrialised Russian North," *Europe–Asia Studies* 64, no. 4 (2012): 645–671.

Bolotova, Alla and Florian Stammler: "How the North Became Home: Attachment to Place among Industrial Migrants in the Murmansk Region of Russia," in Lee Huskey and Chris Southcott, eds., *Migration in the Circumpolar North: Issues and Contexts* (Edmonton: CCI Press, 2010), 193–220.

Bonhomme, Brian: "A Revolution in the Forests? Forest Conservation in Soviet Russia, 1917–1925," *Environmental History* 7, no. 3 (2002): 411–434.

Bonneuil, Christophe and Jean-Baptiste Fressoz: *The Shock of the Anthropocene: The Earth, History and Us*. Translated by David Fernbach (London: Verso Books, 2016).

Bord na Móna: *Fuel Peat in Developing Countries*. World Bank Technical Paper Number 41 (Washington DC: The World Bank, 1985).

Borges, Marcelo J. and Susana B. Torres: "Company Towns: Concepts, Historiography, and Approaches," in Marcelo J. Borges and Susana B. Torres, eds., *Company Towns: Labor, Space, and Power Relations across Time and Continents* (New York: Palgrave Macmillan, 2012), 1–40.

Boscani Leoni, Simona: "Die Debatte um den Torfabbau im 18. Jahrhundert. Die Gebrüder Scheuchzer zwischen Johannes von Muralt und Johann I. Bernoulli," *Jahrbuch für Geschichte des ländlichen Raums* 14 (2017): 81–95.

Bradley, Joseph: *Muzhik and Muscovite: Urbanization in Late Imperial Russia* (Berkeley: University of California Press, 1985).

Brain, Stephen: "The Appeal of Appearing Green: Soviet–American Ideological Competition and Cold War Environmental Diplomacy," *Cold War History* 16, no. 4 (2016): 443–462.

Song of the Forest: Russian Forestry and Stalinist Environmentalism, 1905–1953 (Pittsburgh: University of Pittsburgh Press, 2011).

Braun, Matthias: "Von Menschen und Mikroben: Malaria im Zarenreich und der Sowjetunion in globaler Perspektive," in Martin Aust, ed., *Globalisierung imperial und sozialistisch: Russland und die Sowjetunion in der Globalgeschichte 1851–1991* (Frankfurt: Campus, 2013), 353–372.

Breathnach, Proinnsias: "Engineering and Re-engineering Earth: Industrialized Harvesting of Ireland's Peatlands and Its Aftermath," in Stanley D. Brunn, ed., *Engineering Earth: The Impacts of Megaengineering Projects* (Dordrecht: Springer, 2011), 429–446.

Breyfogle, Nicholas B.: "At the Watershed: 1958 and the Beginnings of Lake Baikal Environmentalism," *Slavonic and East European Review* 93, no. 1 (2015): 147–180.

Bridge, Gavin: "Material Worlds: Natural Resources, Resource Geography and the Material Economy," *Geography Compass* 3, no. 3 (2009): 1217–1244.

Brinegar, Sara C.: *Power and the Politics of Oil in the Soviet South Caucasus: Periphery Unbound, 1920–1929* (London: Bloomsburg Academic, 2024).

Brown, Kate: *Dispatches from Dystopia: Histories of Places Not Yet Forgotten* (Chicago: The University of Chicago Press, 2015).

"Gridded Lives: Why Kazakhstan and Montana are Nearly the Same Place," *The American Historical Review* 106, no. 1 (2001): 17-48.

Manual for Survival: A Chernobyl Guide to the Future (London: Penguin, 2019).

Bruce-Chwatt, Leonard J.: "Malaria Research and Eradication in the USSR: A Review of Soviet Achievements in the Field of Malariology," *Bulletin of the World Health Organization* 21, no. 6 (1959): 737–772.

Bruisch, Katja: "How Wetlands Entered the Transnational Spaces of Late Soviet Environmentalism," in Catherine Evtuhov, Julia Lajus, and David Moon, eds., *Thinking Russia's History Environmentally* (New York: Berghahn, 2023), 201–221.

"Nature Mistaken: Resource-Making, Emotions and the Transformation of Peatlands in the Russian Empire and the Soviet Union," *Environment and History* 26, no. 3 (2020): 359–382.

"The Soviet Village Revisited: Household Farming and the Changing Image of Socialism in the Late Soviet period," *Cahiers du Monde russe* 57, no. 1 (2016): 81–200.

"The State in the Swamps: Territorialization and Ecosystem Engineering in the Western Provinces of the Late Russian Empire," *Zeitschrift für Ostmitteleuropa-Forschung* 68, no. 3 (2019): 345–368.

Bruno, Andy: *The Nature of Soviet Power: An Arctic Environmental History* (Cambridge: Cambridge University Press, 2016).

"Tumbling Snow: Vulnerability to Avalanches in the Soviet North," *Environmental History* 18, no. 4 (2013): 683-709.

"What Does It Mean to Liberate a Land? Toward an Environmental History of the Russian Revolution," in Christopher Read, Peter Waldron, and Adele Lindenmeyr, eds., *Russia's Home Front in War and Revolution, 1914–22*, Book 3: National Disintegration (Bloomington: Slavica Publishers, 2018), 157–177.

Burds, Jeffrey: *Peasant Dreams and Market Politics: Labor Migration and the Russian Village, 1861–1905* (Pittsburgh: University of Pittsburgh Press, 1998).

Burkush, Kateryna: "On the Forest Front: Labour Relations and Seasonal Migration in 1960s–80s," *Labor History* 59, no. 3 (2018): 295–315.

Castree, Noel: "Commodifying what Nature?," *Progress in Human Geography* 27, no. 3 (2003): 273–297.
Chakrabarty, Dipesh: *The Climate of History in a Planetary Age* (Chicago: University of Chicago Press, 2021).
Chernyshova, Natalya: *Soviet Consumer Culture in the Brezhnev Era* (London: Routledge, 2013).
Chu, Pey-Yi: *The Life of Permafrost: A History of Frozen Earth in Russian and Soviet Science* (Toronto: University of Toronto Press, 2021).
Conterio, Johanna: "Controlling Land, Controlling People: Urban Greening and the Territorial Turn in Theories of Urban Planning in the Soviet Union, 1931–1932," *Journal of Urban History* 48, no. 3 (2022): 479–503.
Conze, Susanne: "Women's Work and Emancipation in the Soviet Union, 1941–50," in Melanie Ilič, ed., *Women in the Stalin Era* (Basingstoke: Palgrave, 2001), 216–234.
Coopersmith, Jonathan: *The Electrification of Russia, 1880–1926* (Ithaca: Cornell University Press, 1992).
Corney, Frederick C.: "Rethinking a Great Event: The October Revolution as a Memory Project," *Social Science History* 22, no. 4 (1998): 389–414.
Costlow, Jane: *Heart-Pine Russia: Walking and Writing the Nineteenth Century Forest* (Ithaca: Cornell University Press, 2013).
"Seeing Oil: Isaak Levitan and the Industrial Volga," in Catherine Evtuhov, Julia Lajus and David Moon, eds., *Thinking Russia's History Environmentally* (New York: Berghahn, 2023), 74–97.
Coumel, Laurent: "Building a Soviet Eco-Power while Looking at the Capitalist World: The Rise of Technocratic Environmentalism in Russian Water Controversies, 1957–1989," in Astrid Mignon Kirchhoff and J. R. McNeill, eds., *Nature and the Iron Curtain: Environmental Policy and Social Movements in Communist and Capitalist Countries 1945–1990* (Pittsburgh: University of Pittsburgh Press, 2019), 17–35.
"A Failed Environmental Turn? Khrushchev's Thaw and Nature Protection in Soviet Russia," *The Soviet and Post-Soviet Review* 40 (2013): 167–189.
"La Mechtchera au XXe siècle. L'environnement russe entre ressource(s) et patrimoine," *Siècles* no. 42 (2016), doi: http://siecles.revues.org/2962.
Cresswell, Tim: *Place: A Short Introduction* (Malden, MA: Blackwell Publishing, 2004).
Crippa, M., et al.: *GHG Emissions of All World Countries* (Luxembourg: Publications Office of the European Union, 2023), doi: 10.2760/953322, JRC134504.
Crosby, Alfred W.: *Children of the Sun: A History of Humanity's Unappeasable Appetite for Energy* (New York: W. W. Norton, 2006).
Crowley, Stephen: "Global Cities versus Rustbelt Realities: The Dilemmas of Urban Development in Russia," *Slavic Review* 79, no. 2 (2020): 365–389.
Cummins, Alex: "The Road to NEP, the State Commission for the Electrification of Russia (GOELRO). A Study in Technology, Mobilisation, and Economic Planning," PhD Dissertation, University of Maryland (1988).
van Dam, Petra J. E. M.: "Sinking Peat Bogs: Environmental Change in Holland, 1350–1550," *Environmental History* 6, no. 1 (2001): 32–45.

Davis, Gerald H.: "The Life of Prisoners of War in Russia, 1914–1921," in Samuel R. Williamson Jr. and Peter Pastor, eds., *Essays on World War 1: Origins and Prisoners of War* (New York: Brooklyn College Press, 1983), 163–196.

Demuth, Bathsheba: *Floating Coast: An Environmental History of the Bering Strait* (New York: W. W. Norton, 2019).

Denisova, Liubov: *Rural Women in the Soviet Union and Post-Soviet Russia.* Edited and translated by Irina Mukhina (Abingdon: Routledge, 2010).

Després, Laure and Ksenya Khinchuk: "The Hidden Sector in Soviet Agriculture: A Study of the Military Sovkhozy and Auxiliary Farms," *Soviet Studies* 42, no. 2 (1990): 269–293.

Di Lucia, Lorenzo and Karin Ericsson: "Low-carbon District Heating in Sweden: Examining a Successful Energy Transition," *Energy Research & Social Science* 4 (2014): 10–20.

Di Palma, Vittoria: *Wasteland: A History* (New Haven, CT: Yale University Press, 2014).

Donovan, Victoria: *Chronicles in Stone: Preservation, Patriotism, and Identity in Northwest Russia* (Ithaca: Northern Illinois University Press, 2019).

Doose, Katja: "A Global Problem in a Divided World: Climate Change Research during the Late Cold War, 1972–1991," *Cold War History* 21, no. 4 (2021): 469–489.

Dorondel, Stefan and Anna Olenenko: "In Quest of Development: Territorialization and the Transformation of the Southern Ukrainian Wetlands, 1880–1960," in Stefan Dorondel and Stelu Şerban, eds., *A New Ecological Order: Development and the Transformation of Nature in Eastern Europe* (Pittsburgh: The University of Pittsburgh Press, 2022), 65–86.

Dournel, Sylvain: "De l'exploitation historique de la tourbe dans la vallée de la Somme à la mise en nature des zones humides dans l'agglomération amiénoise: Mutations paysagères et sociales," in J. M. Derex and F. Grégoire, eds., *Histoire économique et sociale de la tourbe et des tourbières* (Cordemais: Æesturia & GHZH, 2009), 211–228.

Dronin, Nikolai and Edward Bellinger: *Climate Dependence and Food Problems in Russia 1900–1990: The Interaction of Climate and Agricultural Policy and Their Effect on Food Problems* (Budapest: CEU Press, 2005).

Dronin, N. M. and J. M. Francis: "Econationalism in Soviet Literature," *Soviet and Post-Soviet Review* 45 (2018): 51–72.

Dvornichenko, V. V. and V. V. Sidorov, "Novaia stoianka na tugolesskikh ozerakh," *Arkheologicheskie Otkrytiia* (1973): 62–63.

Edgerton, David: *The Shock of the Old: Technology and Global History since 1900* (London: Profile Books, 2008, first published 2006).

Elie, Marc: "Formulating the Global Environment: Soviet Soil Scientists and the International Desertification Discussion, 1968–91," *Slavonic and East European Review* 93, no. 1 (2015): 181–204.

Elie, Marc and Klaus Gestwa: "Zwischen Risikogesellschaft und Katastrophenkulturen: Zur Einführung in die Katastrophengeschichte des östlichen Europas," *Jahrbücher für Geschichte Osteuropas* 62, no. 2 (2014): 161–179.

Elliot, Iain F.: *The Soviet Energy Balance: Natural Gas, Other Fossil Fuels, and Alternative Power Sources* (New York: Praeger Publishers, 1974).

Ely, Christopher: *This Meager Nature: Landscape and National Identity in Imperial Russia* (DeKalb: Northern Illinois University Press, 2002).

Erley, Mieka: *On Russian Soil: Myth and Materiality* (Ithaca: Northern Illinois University Press, 2021).

Etkind, Alexander: *Nature's Evil: A Cultural History of Natural Resources.* Translated by Sara Jolly (Cambridge: Polity Press, 2021).

Etkind, Alexander, Yevhenii Poliakov, and Bohdan Shumylovych, "Ukrainian Labor and Siberian Oil in the Late Soviet Empire," *Journal of Soviet and post-Soviet Politics and Society* 6, no. 2 (2020): 241–280.

Evtuhov, Catherine: *Portrait of a Russian Province: Economy, Society, and Civilization in Nineteenth-Century Nizhnii Novgorod* (Pittsburgh: University of Pittsburgh Press, 2011).

Feehan, John and Grace O'Donovan: *The Bogs of Ireland: An Introduction to the Natural, Cultural and Industrial Heritage of Irish Peatlands* (Dublin: University College Dublin Press, 1996).

Feshbach, Murray and Alfred Friendly: *Ecocide in the USSR: Health and Nature under Siege* (London: Aurum Press, 1992).

Filtzer, Donald: "The Contradictions of the Marketless Market: Self-Financing in the Soviet Industrial Enterprise, 1986–90," *Soviet Studies* 43, no. 6 (1991): 989–1009.

Soviet Workers and Late Stalinism: Labour and the Restoration of the Stalinist System after World War II (Cambridge: Cambridge University Press, 2002).

"Women Workers in the Khrushchev Era," in Melanie Ilič, Susan E. Reid, and Lynne Attwood, eds., *Women in the Khrushchev Era* (Basingstoke: Palgrave Macmillan, 2004), 29–51.

Fischer-Kowalski, Marina and Helmut Haberl: "Social Metabolism: A Metrics for Biophysical Growth and Degrowth," in Joan Martínez-Alier and Roldan Muradian, eds., *Handbook of Ecological Economics* (Cheltenham, UK: Edward Elgar, 2015), 100–138.

Fitzpatrick, Sheila: *Stalin's Peasants: Resistance and Survival in the Russian Village after Collectivisation* (New York: Oxford University Press, 1994).

Fluet-Chouinard, Etienne, et al.: "Extensive Global Wetland Loss over the Past Three Centuries," *Nature* 614 (2023): 281–286.

Foster, John Bellamy: "Marx's Theory of Metabolic Rift: Classical Foundations for Environmental Sociology," *American Journal of Sociology* 105, no. 2 (1999): 366–405.

Fraser, Nancy: "Climates of Capital: For a Trans-Environmental Eco-Socialism," *New Left Review* no. 127 (2021): 94–127.

Frey, Felix: *Arktischer Heizraum: Das Energiesystem Kola zwischen regionaler Autarkie und gesamtstaatlicher Verflechtung 1928–1974* (Cologne: Böhlau, 2019).

"Putting Oceans to Work: Tidal Energy in the USA and the USSR, 1930–1970," *History and Technology* 37 no. 4 (2021): 487–504.

Frierson, Cathy A.: *All Russia Is Burning: A Cultural History of Fire and Arson in Late Imperial Russia* (Seattle: University of Washington Press, 2002).

Gatrell, Peter: *Russia's First World War: A Social and Economic History* (Harlow: Pearson Educated Limited, 2005).

Gearey, Benjamin and Ralph Fyfe: "Peatlands as Knowledge Archives," in Aletta Bonn, Tim Allot, Martin Evans, Hans Joosten, and Rob Stoneman, eds., *Peatland Restoration and Ecosystem Services: Science, Policy and Practice* (Cambridge: Cambridge University Press, 2016), 95–113.

Gerasimov, Ilya V.: *Modernism and Public Reform in Late Imperial Russia: Rural Professionals and Self-Organization 1905–1930* (Houndmills: Palgrave Macmillan, 2009).

Gestwa, Klaus: "Das Besitzergreifen von Natur und Gesellschaft im Stalinismus. Enthusiastischer Umgestaltungswille und katastrophischer Fortschritt," *Saeculum* 56, no. 1 (2005): 105–138.

Die Stalinschen Großbauten des Kommunismus: Sowjetische Technik- und Umweltgeschichte, 1948–1967 (Munich: Oldenbourg, 2010).

Goldman, Marshall I.: *The Spoils of Progress: Environmental Pollution in the Soviet Union* (Cambridge, MA: MIT Press, 1972).

Gradskova, Yulia: "Soviet People with Female Bodies. Performing Beauty and Maternity in Soviet Russian in the mid 1930–1960s," PhD dissertation, Stockholm University (2007).

Gregory, Paul R.: *Russian National Income, 1885–1913* (Cambridge: Cambridge University Press 1982).

Gritsenko, Daria and Hilma Salonen: "A Local Perspective on Renewable Energy Development in the Russian Artic," *Elementa: Science of the Anthropocene* 8, no. 1 (2020): article number 441, doi: https://doi.org/10.1525/elementa.441.

Günther A., et al.: "Prompt Rewetting of Drained Peatlands Reduces Climate Warming Despite Methane Emissions," *Nature Communications* 11 (2020): article number 1644, doi: https://doi.org/10.1038/s41467-020-15499-z.

Gustafson, Thane: *The Bridge: Natural Gas in a Redivided Europe* (Cambridge, MA: Harvard University Press, 2020).

Halperin, Charles J.: "Soviet Historiography on Russia and the Mongols," *The Russian Review* 41, no. 3 (1982): 306–322.

Hanson, Philip: *The Rise and Fall of the Soviet Economy: An Economic History of the USSR from 1945* (Oxon: Routledge, 2003).

Harrison, Mark: "The Soviet Union after 1945: Economic Recovery and Political Repression," *Past and Present* 210, Supplement 6 (2011): 103–120.

Harrison, Mark and Byung-Yeon Kim: "Plans, Prices and Corruption: The Soviet Firm under Partial Centralization, 1930 to 1990," *The Journal of Economic History* 66, no. 1 (2006): 1–41.

Harvey, David: *Justice, Nature and the Geography of Difference* (Cambridge, MA: Blackwell Publishers, 1996).

Hasenöhrl, Ute and Patrick Kupper: "Historicizing Renewables: Issues and Challenges," *History and Technology* 37, no. 4 (2021): 397–410.

Haumann, Heiko: *Beginn der Planwirtschaft: Elektrifizierung, Wirtschaftsplanung und gesellschaftliche Entwicklung Sowjetrusslands 1917–1921* (Düsseldorf: Bertelsmann, 1974).

Healey, Dan: "Comrades, Queers, and 'Oddballs': Sodomy, Masculinity, and Gendered Violence in Leningrad Province of the 1950s," *Journal of the History of Sexuality* 21, no. 3 (2012): 496–522.

Hedlund, Stefan: *Private Agriculture in the Soviet Union* (London: Routledge, 1989).

Hickel, Jason: "Quantifying National Responsibility for Climate Breakdown: An Equality-Based Attribution Approach for Carbon Dioxide Emissions in Excess of the Planetary Boundary," *The Lancet Planetary Health* 4, no. 9 (2020): e399–e404.

Hirschler, Oliver and Bernhard Osterburg: "Peat Extraction, Trade and Use in Europe: A Material Flow Analysis," *Mires and Peat* 28, no. 24 (2022), doi: 10.19189/MaP.2021.SNPG.StA.2315.

Hoffmann, David L.: "Mothers in the Motherland: Stalinist Pronatalism in Its Pan-European Context," *Journal of Social History* 34, no. 1 (2000): 35–54.

Peasant Metropolis: Social Identities in Moscow, 1929–1941 (Ithaca, NY: Cornell University Press, 1994).

Högselius, Per: *Red Gas: Russia and the Origins of European Energy Dependence* (New York: Palgrave Macmillan, 2013).

Holquist, Peter: *Making War, Forging Revolution: Russia's Continuum of Crisis 1914–1921* (Cambridge, MA: Harvard University Press, 2002).

Hughes, Thomas P.: *Networks of Power: Electrification in Western Society, 1880–1930* (Baltimore: The Johns Hopkins University Press, 1983).

Humphrey, Caroline: *The Unmaking of Soviet Life: Everyday Economies after Socialism* (Ithaca: Cornell University Press, 2002).

Iacob, Bogdan C.: "Malariology and Decolonization: Eastern European Experts from the League of Nations to the World Health Organization," *Journal of Global History* 17, no. 2 (2022): 233–253.

Ianitskii, O. N.: "Pozhary v 2010 g. v Rossii: Ėkosotsiologicheskii analiz," *Sotsiologicheskie issledovaniia* no. 3 (2011): 3–12.

Ilič, Melanie: "'Equal Pay for Equal Work': Women's Wages in Soviet Russia," in Melanie Ilič, ed., *The Palgrave Handbook of Women and Gender in Twentieth-Century Russia and the Soviet Union* (London: Palgrave Macmillan, 2018), 101–115.

Ingold, Tim: *Lines: A Brief History* (London: Routledge, 2007).

Ioffe, Grigory and Zhanna Zayonchkovskaya: "Spatial Shifts in the Population of Moscow Region," *Eurasian Geography and Economics* 52, no. 4 (2011): 543–566.

IPCC: *2006 IPCC Guidelines for National Greenhouse Gas Inventories*, Vol. 2, Energy (Hayama: Institute for Global Environmental Strategies, 2006).

Iriarte-Goñi, Iñaki and María-Isabel Ayuda: "Not Only Subterranean Forests: Wood Consumption and Economic Development in Britain (1850–1938)," *Ecological Economics* 77 (2012): 176–184.

Irvine, Richard G. D.: *An Anthropology of Deep Time: Geological Temporality and Social Life* (Cambridge: Cambridge University Press, 2020).

Javanshir, Nima, et al.: "Abandoning Peat in a City District Heat System with Wind Power, Heat Pumps, and Heat Storage," *Energy Reports* 8 (2022): 3051–3062.

Jones, Christopher F.: "Petromyopia: Oil and the Energy Humanities," *Humanities* 5, no. 2 (2016), doi: https://doi.org/10.3390/h5020036.

Routes of Power: Energy and Modern America (Cambridge, MA: Harvard University Press, 2014).

Jones, Polly: *Myth, Memory, Trauma: Rethinking the Stalinist Past in the Soviet Union, 1953–70* (New Haven: Yale University Press, 2013).

Jones, Ryan Tucker: *Red Leviathan: The Secret History of Soviet Whaling* (Chicago: The University of Chicago Press, 2022).

Joosten, Hans: "Human Impacts: Farming, Fire, Forestry and Fuel," in Edward Maltby and Tom Barker, eds., *The Wetlands Handbook* (Chichester: Wiley-Blackwell, 2009), 689–718.

"The International Peat Society: Fossil or Renewable? An Analysis of the IPS Stand towards Peat Renewability and Climate Change," *IMCG Newsletter* no. 2 (2007): 4–19.

Joosten, Hans and Donal Clarke: *Wise Use of Mires and Peatlands: Background and Principles including a Framework for Decision-Making* (Devon: International Mire Conservation Group & International Peat Society, 2002).

Joosten, Hans, Franziska Tanneberger, and Asbjørn Moen, eds., *Mires and Peatlands of Europe: Status, Distribution and Conservation* (Stuttgart: Schweizerbart Science Publishers, 2017).

Josephson, Paul: "'Projects of the Century' in Soviet History: Large-Scale Technologies from Lenin to Gorbachev," *Technology and Culture* 36, no. 3 (1995): 519–559.

Kander, Astrid, Paolo Malanima, and Paul Warde: *Power to the People: Energy in Europe over the Last Five Centuries* (Princeton: Princeton University Press, 2013).

Kasatkina, Aleksandra: "K poėtike i praktike modernogo urbanizma: 'Gorod-sad' kak tekhnologiia sotsial'noi integratsii v Obninske 1960-kh godov," *Laboratorium: Zhurnal sotsial'nykh issledovanii* 14, no. 1 (2022): 30–59.

Keating, Jennifer: *On Arid Ground: Political Ecologies of Empire in Tsarist Central Asia* (Oxford: Oxford University Press, 2022).

Keller, Shoshana: "The Puzzle of Manual Harvest in Uzbekistan: Economics, Status and Labour in the Khrushchev Era," *Central Asian Survey* 34, no. 3 (2015): 296–309.

Keller, Tait: "The Ecological Edges of Belligerency: Toward a Global Environmental History of the First World War," *Annales. Histoire, Sciences Sociales (English Edition)* 71, no. 1 (2016): 61–78.

Kelman Wieder, R., Merritt R. Turetsky, and Melanie A. Vile: "Peat as an Archive of Atmospheric, Climatic and Environmental Conditions," in Edward Maltby and Tom Barker, eds., *The Wetlands Handbook* (Chichester: Wiley-Blackwell, 2009), 96–112.

Kerbikov, Mikhail D.: "Torfushki: Vzgliad na istoriiu zabytogo Truda," in K. A. Pakhaliuk, ed., *Georgievskie chteniia: Sbornik trudov po voennoi istorii otechestva* (Moscow: Iauza, 2021), 523–536.

Kessler, Gijs: "The Peasant and the Town. Rural–Urban Migration in the Soviet Union, 1929–40," PhD dissertation, European University Institute (2001).

"The Rural–Urban Nexus in Russian Labour History, 1860s–1930s: Suggestions for a Global, Comparative History," in Marcel van der Linden and Prabhu P. Mohapatra, eds., *Labour Matters: Towards Global Histories: Studies in Honour of Sabyasachi Bhattacharya* (New Delhi: Tulika Books, 2009), 207–225.

Khanin, G. I.: "The 1950s: The Triumph of the Soviet Economy," *Europe–Asia Studies* 55, no. 8 (2003): 1187–1211.

Kirsch, Scott and Don Mitchell: "The Nature of Things: Dead Labor, Nonhuman Actors, and the Persistence of Marxism," *Antipode: A Radical Journal of Geography* 36, no. 4 (2004): 687–705.

Kochetkova, Elena: *The Green Power of Socialism: Wood, Forest, and the Making of Soviet Industrially Embedded Ecology* (Cambridge, MA: The MIT Press, 2024).

Kochetkova, Elena and Anna Petrova: "Abandoned, But Not Forgotten Heritage: Former Industrial Enterprises in Cultural and Urban Russian Landscapes," *Heritage & Society* 15, no. 3 (2022): 277–298.

Kochetkova, Elena: "Milk and Milk Packaging in the Soviet Union," *Russian History* 46, no. 1 (2019): 29–52.

Kopenkina, L. V.: *Istoriia torfianogo dela v Rossii* (Tver': TGTU, 2015).

"Torfianaia otrasl' v gody Velikoi Otechestvennoi Voiny (k 70-letiiu Velikoi Pobedy)," *Trudy Instorfa* 13, no. 66 (2016): 50–56.

Korchmina, Elena: "Conceptualising the Notion of 'Profit' among the Russian Nobility (Second Half of the 18th–First Half of the 19th Centuries)," *Quaestio Rossica* 6 no. 4 (2018): 1144–1159.

Kotsonis, Yanni: *Making Peasants Backward: Agricultural Cooperatives and the Agrarian Question in Russia, 1861–1914* (New York: St. Martin's Press, 1999).

Kragh, Martin: "The Soviet Enterprise: What Have We Learned from the Archives?," *Enterprise & Society* 14, no. 2 (2013): 360–394.

Kreyling, J., et al: "Rewetting Does Not Return Drained Fen Peatlands to Their Old Selves," *Nature Communications* 12 (2021): 5693, doi: https://doi.org/10.1038/s41467-021-25619-y.

Kulikov, Volodymyr: "Industrialisation and Transformation of the Landscape in the Donbass from the late Nineteenth to the early Twentieth Century," in Martin Zückert and Heidi Hein-Kirchner, eds., *Migration and Landscape Transformation: Changes in Central and Eastern Europe in the 19th and 20th Century* (Göttingen: Vandenhoeck & Rupprecht, 2016), 57–81.

"Necessity or Luxury? Welfare Work in the Company Towns of the Russian Empire," *Jahrbuch für Wirtschaftsgeschichte* 60, no. 2 (2019): 449–472.

Kulikov, Volodymyr and Martin Kragh: "Big Business in the Russian Empire: A European Perspective," *Business History* 61, no. 2 (2019): 299–321.

Kunnas, Jan and Timo Myllyntaus: "Postponed Leap in Carbon Dioxide Emissions: The Impact of Energy Efficiency, Fuel Choices and Industrial Structure on the Finnish Economy, 1800–2005," *Global Environment* 2, no. 3 (2009): 154–189.

van Laak, Dirk: "Infra-Strukturgeschichte," *Geschichte und Gesellschaft* 27, no. 3 (2001): 367–393.

"Planung: Geschichte und Gegenwart des Vorgriffs auf die Zukunft," *Geschichte und Gesellschaft* 34, no. 3 (2008): 305–326.

Lapierre, Brian: "Making Hooliganism on a Mass Scale: The Campaign against Petty Hooliganism in the Soviet Union, 1956–1964," *Cahiers du Monde russe* 47, no. 1/2 (2006): 349–375.

LeCain, Timothy J.: *The Matter of History: How Things Create the Past* (Cambridge: Cambridge University Press, 2017).

Leifeld, Jens, et al.: "Intact and Managed Peatland Soils as a Source and Sink of GHGs from 1850 to 2100," *Nature Climate Change* 9 (2019): 945–947.

Leonard, Carol S.: *Agrarian Reform in Russia: The Road from Serfdom* (Cambridge: Cambridge University Press, 2011).

Lévesque, Jean: "'Into the Grey Zone': Sham Peasants and the Limits of the Kolkhoz Order in the Post-War Russian Village, 1945–1953," in Juliane Fürst, ed., *Late Stalinist Russia: Society between Reconstruction and Reinvention* (London: Routledge, 2006), 103–119.

Lochmann, Bodo: "Vom GOELRO-Plan zum Energieprogramm. Zur Entwicklung der Elektroenergiewirtschaft der UdSSR seit den zwanziger Jahren," *Zeitschrift für Geschichtswissenschaft* 35, no. 10 (1987): 872–883.

Lupanova, Evgeniia: *Istoriia zakreposhcheniia prirodnogo resursa: Lesnoe khoziaistvo v Rossii 1696–1802 gg.* (Saint Petersburg: EUSP Press, 2017).

Maier, C. S.: "Between Taylorism and Technocracy: European Ideologies and the Vision of Industrial Productivity in the 1920s," *Journal of Contemporary History* 5, no. 2 (1970): 27–61.

Malek, Corinna: "Torfenergie und ihre Folgen in Bayern," in *Kulturerbe Energie: Zeugnisse der Energiegewinnung und Energienutzung als Kulturlandschaftselemente entdecken* (Bonn: Bund Heimat and Umwelt, 2015), 54–64.

Malm, Andreas: *Fossil Capital: The Rise of Steam Power and the Roots of Global Warming* (Brooklyn, NY: Verso Books, 2016).

Maltby, Edward: "The Changing Wetland Paradigm," in Edward Maltby and Tom Barker, eds., *The Wetlands Handbook* (Chichester: Wiley-Blackwell, 2009), 3–42.

Maltby, Edward and Tom Barker, eds., *The Wetlands Handbook* (Chichester: Wiley-Blackwell, 2009).

Maltby, E. and M. C. F. Proctor: "Peatlands: Their Nature and Role in the Biosphere," in Eino Lappalainen, ed., *Global Peat Resources* (Jyskä: International Peat Society, 1996), 11–19.

Markevich, Andrei: "Finding Additional Income: Subsidiary Agriculture of Soviet Urban Households, 1941–1964," in Donald Filtzer, Wendy Z. Goldman, Gijs Kessler, and Simon Pirani, eds., *A Dream Deferred: New Studies in Russian and Soviet Labour History* (Bern: Peter Lang, 2008), 385–415.

Markevich, Andrei and Mark Harrison: "Great War, Civil War, and Recovery: Russia's National Income, 1913 to 1928," *The Journal of Economic History* 71, no. 3 (2011): 672–703.

Markevich, Andrei and Steven Nafziger: "State and Market in Russian Industrialization, 1870–2010," in Kevin Hjortshøj O'Rourke and Jeffrey

Gale Williamson, eds., *The Spread of Modern Industry to the Periphery since 1871* (Oxford: Oxford University Press, 2017), 33–62.
Markevich, Andrei and Ekaterina Zhuravskaya: "The Economic Effects of the Abolition of Serfdom: Evidence from the Russian Empire," *The American Economic Review* 108, no. 4-5 (2018): 1074–1117.
Markov, V. I.: "Periody razvitiia torfianoi promyshlennosti Rossii," *Trudy Instorfa* 6, no. 59 (2012): 10–21.
Matthews, H. Damon: "Quantifying Historical Carbon and Climate Debts among Nations," *Nature Climate Change* 6 (2016): 60–64.
McAuley, Alastair: *Women's Work and Wages in the Soviet Union* (London: George Allen & Unwin, 1981).
McEvoy, Arthur F.: "Working Environments: An Ecological Approach to Industrial Health and Safety," *Technology and Culture* 36, no. 2 (1995): S145–S173.
McNeill, John R.: "Cheap Energy and Ecological Teleconnections of the Industrial Revolution, 1780–1920," *Environmental History* 24, no. 3 (2019): 492–503.
Mosquito Empires: Ecology and War in the Greater Caribbean, 1620–1914 (Cambridge: Cambridge University Press, 2010).
"The State of the Field of Environmental History," *Annual Review of Environment and Resources* 35 (2010): 345–374.
McNeill, John R. and Peter Engelke: *The Great Acceleration: An Environmental History of the Anthropocene since 1945* (Cambridge, MA: Belknap Press of Harvard University Press, 2014).
Meerovich, Mark: *Gradostroitel'naia politika v SSSR 1917–1929: Ot goroda-sada k vedomstvennomu rabochemu poselku*, 2nd ed. (Moscow: NLO, 2018).
Minayeva, Tatiana and Andrey Sirin, eds., *A Quick Scan of Peatlands in Central and Eastern Europe* (Wageningen: Wetlands International, 2009).
Minayeva, T., A. A. Sirin, and G. B. Stracher: "The Peat Fires of Russia," in Glenn B. Stracher, Anupma Prakash, and Ellina V. Sokol, eds., *Coal and Peat Fires: A Global Perspective*, Vol. 2, Photographs and Multimedia Tours (Amsterdam: Elsevier, 2013), 375–394.
Mitchell, Timothy: *Carbon Democracy: Political Power in the Age of Oil* (London: Verso, 2013).
Rule of Experts: Egypt, Techno-Politics, Modernity (Berkeley: University of California Press, 2002).
Mitman, Gregg: "Living in a Material World," *The Journal of American History* 100, no. 1 (2013): 128–130.
Montaño, Diana J.: "Ladrones de Luz: Policing Electricity in Mexico City, 1901–1918," *Hispanic American Historical Review* 101, no. 1 (2011): 35–72.
Moon, David: *The Plough that Broke the Steppes: Agriculture and Environment on Russia's Grasslands, 1700–1914* (Oxford: Oxford University Press, 2013).
The Russian Peasantry: The World the Peasants Made, 1600 – 1930 (London and New York: Routledge, 1999)

Moon, David, Nicholas B. Breyfogle, and Alexandra Bekasova, eds., *Place and Nature: Essays in Russian Environmental History* (Cambridgeshire: White Horse Press, 2021).

Morris, Jeremy: *Everyday Post-Socialism: Working-Class Communities in the Russian Margins* (London: Palgrave Macmillan, 2016).

Moritz, Verena: "Die österreichisch-ungarischen Kriegsgefangenen in der russischen Wirtschaft (1914 bis Oktober 1917)," *zeitgeschichte* 25, no. 11–12 (1998): 380–389.

Moser, Nat: *Oil and the Economy of Russia: From the Late-Tsarist to the Post-Soviet Period* (London: Routledge, 2018).

Moskoff, William: *Labour and Leisure in the Soviet Union: The Conflict between Public and Private Decision-Making in a Planned Economy* (London: Macmillan, 1984).

Nachtigall, Reinhard: *Rußland und seine österreichisch-ungarischen Kriegsgefangenen (1914–1918)* (Remshalden: Verlag Bernhard Albert Greiner, 2003).

Nash, Catherine: "Environmental History, Philosophy and Difference," *Journal of Historical Geography* 26, no. 1 (2000): 23–27.

Nefedova, T. G.: "Khronika i prichiny goriachego leta 2010," *Znanie – Sila*, no. 6 (2011): 49–57.

"Rossiiskaia periferiia kak sotsial'no-ėkonomicheskii fenomen," *Regional'nye issledovaniia* 20, no. 5 (2008): 14–30.

Nekrasov, Viacheslav: "Decision-Making in the Soviet Energy Sector in Post-Stalinist Times: The Failure of Khrushchev's Economic Modernization Strategy," in Jeronim Perovič, ed., *Cold War Energy: A Transnational History of Soviet Oil and Gas* (Cham: Palgrave Macmillan, 2017), 165–199.

Nikiforova, Natalia: "'Useless Fossils', Precious Waste, and Streams of Energy: Soviet Electrification and Natural Resources for the Socialist Future (1920s–1930s)," *Technology and Language* 4, no. 2 (2023): 72–87.

Nikishova, Elena V.: *El'nikovtsy*, Vol. 3 (Saransk: K. Shapkarin, 2019).

Nikkilä, Lea-Elina: "Hydropeat: An Old Harvesting Method," in Riita Korhonen, Leila Korpela, and Sakari Sarkkolo, eds., *Finland–Fenland: Research and Sustainable Utilisation of Mires and Peat* (Maahenki: Finnish Peatland Society, 2008), 106–109.

Nixon, Rob: "The Anthropocene: The Promise and Pitfalls of an Epochal Idea," in Gregg Mitman, Marco Armiero, and Robert S. Emmett, eds., *Future Remains: A Cabinet of Curiosities for the Anthropocene* (Chicago: The University of Chicago Press, 2018), 1–18.

Nove, Alec: *An Economic History of the USSR, 1917–1991*, 3rd ed. (London: Penguin, 1992).

Oberländer, Alexandra: "Cushy Work, Backbreaking Leisure: Late Soviet Work Ethics Reconsidered," *Kritika: Explorations in Russian and Eurasian History* 18 no. 3 (2017): 569–590.

"Hatching Money: The Political Economy of Eggs in the 1960s," *Cahiers du Monde russe* 61, no. 1–2 (2020): 231–256.

Obertreis, Julia: "Von der Naturbeherrschung zum Ökozid? Aktuelle Fragen einer Umweltzeitgeschichte Ost- und Ostmitteleuropas," *Zeithistorische Forschungen* 9, no. 1 (2012): 115–122.

O'Gorman, Emily and Andrea Gaynor: "More-Than-Human Histories," *Environmental History* 25, no. 4 (2020): 711–735.

Oldfield, Jonathan: "Imagining Climates Past, Present and Future: Soviet Contributions to the Science of Anthropogenic Climate Change, 1953–1991," *Journal of Historical Geography* 60 (2018): 41–51.

Oldfield, Jonathan and Denis J. B. Shaw: "V. I. Vernadskii and the Development of Biogeochemical Understandings of the Biosphere, c.1880s–1968," *British Journal for the History of Science* 46, no. 2 (2013): 287–310.

Olenenko, Anna: "'Novoe nashe more – novoe nashe gore': Konflikt mezhdu ukrainskim i sovetskim v bor'be za konstruirovanie landshafta Nizhnego Podneprov'ia," *Ab Imperio* no. 1 (2019): 125–152.

Oushakine, Serguei A.: "'Against the Cult of Things': On Soviet Productivism, Storage Economy, and Commodities with No Destination," *The Russian Review* 73, no. 2 (2014): 198–236.

Palvinskaya, Zinaida: "How War Affects Grassroots Environmental Protest," *Environment & Rights* 85 (October 2022): 36–41.

Patel, Raj and J. W. Moore: *A History of the World in Seven Cheap Things: A Guide to Capitalism, Nature, and the Future of the Planet* (London: Verso Books, 2018).

Peatland Atlas: Facts and Figures about Wet Climate Guardians (Berlin: Heinrich-Böll-Stiftung et al., 2023).

Perović, Jeronim, ed., *Cold War Energy: A Transnational History of Soviet Oil and Gas* (Cham: Palgrave Macmillan, 2017).

Rohstoffmacht Russland: Eine globale Energiegeschichte (Cologne: Böhlau, 2022).

Peterson, Maya K.: *Pipe Dreams: Water and Empire in Central Asia's Aral Sea Basin* (Cambridge: Cambridge University Press, 2019).

Pianciola, Niccolò: "Cossacks and Sturgeons: Fisheries, Colonization and Science around the Aral Sea (1873–1906)," *Journal of the Economic and Social History of the Orient* 62, no. 4 (2019): 626–673.

Pickering Antonova, Katherine: "Discovering Russian Regions: Fruits of the Archival Turn in Imperial Russian History," *Kritika: Explorations in Russian and Eurasian History* 21, no. 4 (2020): 877–892.

"Regional'naia paradigma ėkonomicheskogo razvitiia v Rossii XIX veka," in Ekaterina Boltunova and Willard Sunderland, eds., *Regiony Rossiiskoi imperii: Identichnost', reprezentatsiia, (na)znachenie* (Moscow: NLO, 2021), 52–74.

Pirani, Simon: *The Russian Revolution in Retreat, 1920–24: Soviet Workers and the New Communist Elite* (London: Routledge, 2008).

Pohl, Michaela: "The 'Planet of One Hundred Languages': Ethnic Relations and Soviet Identity in the Virgin Lands," in Nicholas B. Breyfogle, Abby Schrader, and Willard Sunderland, eds., *Peopling the Russian Periphery: Borderland Colonization in Eurasian History* (London: Routledge, 2007), 238–261.

Pospiszyl, Michał: "The Fifth Element: The Enlightenment and the Draining of Eastern Europe," *Environmental History* 28, no. 2 (2023): 361–388.

Potkina, I. V.: *Na olimpe delovogo uspekha: Nikol'skaia manufaktura Morozovykh* (Moscow: Izdatel'stvo Glavarkhiva Moskvy, 2004).

Pravilova, Ekaterina: *A Public Empire: Property and the Quest for the Common Good in Imperial Russia* (Princeton, NJ: Princeton University Press, 2014).
Proulx, Annie: *Fen, Bog & Swamp: A Short History of Peatland Destruction and Its Role in the Climate Crisis* (London: 4th Estate, 2022).
Rachamimov, Alon: *POWs and the Great War: Captivity on the Eastern Front* (Oxford: Berg, 2002).
Rakhimzianov, Bulat R.: "Meshchera as a Point of Political Interaction between Muscovy and the Tatar World," *Russian History* 43, no. 3–4 (2016): 373–393.
Rehschuh, Felix: *Aufstieg zur Energiemacht: Der sowjetische Weg ins Erdölzeitalter 1930er bis 1950er Jahre* (Cologne: Böhlau, 2018).
Reid, Susan A.: "Building Utopia in the Back Yard: Housing Administration, Participatory Government, and the Cultivation of Socialist Community," in Karl Schlögel, ed., *Mastering Russian Spaces: Raum und Raumbewältigung als Probleme der russischen Geschichte* (Munich: De Gruyter Oldenbourg, 2011), 149–185.
"Khrushchev Modern: Agency and Modernisation in the Soviet Home," *Cahiers du Monde russe* 47, no. 1–2 (2006): 227–268.
Resheniia partii i pravitel'stva po khoziaistvennym voprosam: Sbornik dokumentov. Tom 11 (Moscow: Politizdat, 1977).
Richardson, Tanya: "Interspecies Relations in the Midst of the Russia–Ukraine War," *Current History* 122, no. 847 (2023): 301–307.
Richardson, Tanya and Gisa Weszkalnys: "Introduction: Resource Materialities," *Anthropological Quarterly* 87, no. 1 (2014): 5–30.
Rindzevičiūtė, Eglė: *The Power of Systems: How Policy Sciences Opened Up the Cold War World* (Ithaca, NY: Cornell University Press, 2017).
Rispoli, Giulia and Jacques Grinevald: "Vladimir Vernadsky and the Co-evolution of the Biosphere, the Noosphere, and the Technosphere," *Technosphere Magazine*, 20 June 2018, https://technosphere-magazine.hkw.de/p/Vladimir-Vernadsky-and-the-Co-evolution-of-the-Biosphere-the-Noosphere-and-the-Technosphere-nuJGbW9KPxrREPxXxz95hr.
Rodionov, N. A.: "Mobilizatsiia torfianykh resursov Urala na nuzhdy voennogo proizvodstva (1941–1945 gg.)," *Istoricheskie, filosofskie, politicheskie i iuridicheskie nauki, kul'turologiia i iskusstvovedenie. Voprosy teorii i praktiki* 1, no. 1 (2015): 168–172.
"Torfopromyshlennoe nasledie Urala XX veka: Problemy izucheniia i sokhraneniia," *Bakuninskie chteniia: Sektsiia 5. Istoriia nauki i tekhniki* (2019): 344–355.
"Voprosy ispol'zovaniia trudovykh resursov Mordovskoi ASSR v torfodobyvaiushchei promyshlennosti Udmurtskoi Respubliki (1940–1960-e gody)," *Ėkonomicheskaia Istoriia* 17, no. 2 (2012): 53–64.
Roe, Alan D.: *Into Russian Nature: Tourism, Environmental Protection, and National Parks in the Twentieth Century* (New York: Oxford University Press, 2020).
Rolf, Malte: "Importing the 'Spatial Turn' to Russia: Recent Studies on the Spatialization of Russian History," *Kritika: Explorations in Russia and Eurasian History* 11, no. 2 (2010): 359–380.

Russ, Daniela: "Energetika: Gleb Krzhizhanovskii's Conception of the Nature–Society Metabolism," *Historical Materialism* 29, no. 2 (2021): 188–218.

"'Socialism Is Not Just Built for a Hundred Years': Renewable Energy and Planetary Thought in the Early Soviet Union (1917–1945)," *Contemporary European History* 31, no. 4 (2022): 491–508.

"Speaking for the 'World Power Economy': Electricity, Energo-Materialist Economics, and the World Energy Council (1924–78)," *Journal of Global History* 15, no. 2 (2020): 311–329.

Ruuskanen, Esa: "The Emergence of Baltic *Moorkultur*: Visions of Scientific–Technological Mastery of Peatlands in the Age of Great Social Change, 1850–1914," *History and Technology* 34, no. 3–4 (2018): 213–234.

"Valuing Peatlands and Wetlands: Mires in Natural Resource and Land Use Policies in the Nordic Countries from the Late Eighteenth Century to the Present Day," in Gordon Winter and Andreas Dix, eds., *Trading Environments: Frontiers, Commercial Knowledge and Environmental Transformation, 1820–1990* (New York: Routledge, 2016), 118–136.

Sack, Robert David: "The Geographic Problematic: Empirical issues," *Norsk Geografisk Tidsskrift: Norwegian Journal of Geography* 55, no. 3 (2001): 107–116.

Saito, Kohei: *Marx in the Anthropocene: Towards the Idea of Degrowth Communism* (Cambridge: Cambridge University Press, 2022).

Sanchez-Sibony, Oscar: "Global Money and Bolshevik Authority: The NEP as the First Socialist Project," *Slavic Review* 78, no. 3 (2019): 694–716.

Saunders, David: "Regional Diversity in the Later Russian Empire," *Transactions of the Royal Historical Society* 10 (2000): 143–163.

Schaeffer Conroy, Mary: "Malaria in Late Tsarist Russia," *Bulletin of the History of Medicine* 56, no. 1 (1982): 41–55.

Schattenberg, Susanne: *Stalins Ingenieure: Lebenswelten zwischen Technik und Terror in den 1930er Jahren* (Munich: Oldenbourg, 2002).

Schaupp, Simon: *Stoffwechselpolitik: Arbeit, Natur und die Zukunft des Planeten* (Berlin: Suhrkamp, 2024).

Schenk, Frithjof Benjamin: "Das Paradigma des Raumes in der Osteuropäischen Geschichte," *zeitenblicke: Online Journal für die Geschichtswissenschaften* 6, no. 2 (2007), www.zeitenblicke.de/2007/2/schenk.

Russlands Fahrt in die Moderne: Mobilität und sozialer Raum im Eisenbahnzeitalter (Stuttgart: Steiner, 2014).

Schlögel, Karl, ed., *Mastering Russian Spaces: Raum und Raumbewältigung als Probleme der russischen Geschichte* (Munich: Oldenbourg, 2011).

Schlüter, Mogens: "The Use of Peat in Danish Glassworks, 1825–1945," *Journal of Glass Studies* 30 (1988): 94–101.

Scott, James C.: *Seeing Like a State: How Certain Schemes to Improve the Human Condition Have Failed* (New Haven, CT: Yale University Press, 1998).

Seow, Victor: *Carbon Technocracy: Energy Regimes in Modern East Asia* (Chicago: The University of Chicago Press, 2021).

Shimkin, Demitri B.: *The Soviet Mineral–Fuels Industries, 1928–1958: A Statistical Survey* (Washington, DC: US Department of Commerce, 1962).

Sidorov, V. V.: "Stoianki na ozere Sviatom u Shatury," *Sovetskaia Arkheologiia* no. 3 (1975): 107–117.

Sieferle, Rolf Peter: *The Subterranean Forest: Energy Systems and the Industrial Revolution*, translated by Michael P. Osman (Cambridge: The White Horse Press, 2001).

Siegelbaum, Lewis H. and Leslie Page Moch: *Broad Is My Native Land: Repertoires and Regimes of Migration in Russia's Twentieth Century* (Ithaca, NY: Cornell University Press, 2014).

Simonov, S. N.: *Razvitie ėlektroėnergetiki Rossiiskoi imperii: Predystoriia GOĖLRO* (Moscow: Universitet Dmitriia Pozharskogo, 2018).

Sirin, A. A. and T. Iu. Minaeva, eds., *Torfianye bolota Rossii: K analizu otraslevoi informatsii* (Moscow: Geos, 2001).

Sirin, A. A., et al.: "Addressing Peatland Rewetting in Russian Federation Climate Reporting," *Land* 10, no. 11, 1200 (2021): 1–17, doi: https://doi.org/10.3390/land10111200.

"Mapping of Peatlands in the Moscow Oblast Based on High-Resolution Remote Sensing Data," *Contemporary Problems of Ecology* 7, no. 1 (2014): 808–814.

"Multispectral Satellite Based Monitoring of Land Cover Change and Associated Fire Reduction after Large-scale Peatland Rewetting following the 2010 Peat Fires in Moscow Region (Russia)," *Ecological Engineering* 158 (2020): 106044, doi: https://doi.org/10.1016/j.ecoleng.2020.106044.

"Russian Federation," in Hans Joosten, Franziska Tanneberger, and Asbjørn Moen, eds., *Mires and Peatlands of Europe: Status, Distribution and Conservation* (Stuttgart: Schweizerbart Science Publishers, 2017), 591–616.

Smil, Vaclav: *Energy and Civilisation: A History* (Cambridge, MA: MIT Press, 2017).

Smith, Alexandra: "The Enigma of Mikhail Prishvin: Prishvin's Pre-Soviet and Soviet Diaries," *New Zealand Slavonic Journal* (2000): 79–92.

Smith, Alison K.: "Lime and Limestone in Eighteenth-Century Russia," in Matthew P. Romaniello, Alison K. Smith, and Tricia Starks, eds., *The Life Cycle of Russian Things: From Fish Guts to Fabergé, 1600–Present* (London: Bloomsbury Academic, 2022), 33–49.

"A Microhistory of the Global Empire of Cotton: Ivanovo, the 'Russian Manchester'," *Past and Present* 244, no. 1 (2019): 163–193.

Smith, Jenny Leigh: *Works in Progress: Plans and Realities on Soviet Farms, 1930–1963* (New Haven: Yale University Press, 2014).

Smith, Mark B.: "Faded Red Paradise: Welfare and the Soviet City after 1953," *Contemporary European History* 24, no. 4 (2015): 597–615.

Smith, S. A.: *Russia in Revolution: An Empire in Crisis, 1890 to 1928* (Oxford: Oxford University Press, 2017).

Smith-Peter, Susan: "How to Write a Region: Local and Regional Historiography," *Kritika: Explorations in Russian and Eurasian History* 5, no. 3 (2004): 527–542.

Snowden, Frank M.: "Mosquitoes, Quinine and the Socialism of Italian Women 1900–1914," *Past & Present* no. 178 (2003): 176–209.

Stanziani, Alessandro: *After Oriental Despotism: Eurasian Growth in a Global Perspective* (London: Bloomsbury Academic, 2014).

Stites, Richard: *Revolutionary Dreams: Utopian Vision and Experimental Life in the Russian Revolution* (New York: Oxford University Press, 1989).

Tagirova, Nailya: "Mapping the Empire's Economic Regions from the Nineteenth to the Early Twentieth Century," in Jane Burbank, Mark von Hagen, and Anatolyi Remnev, eds., *Russian Empire: Space, People, Power, 1700–1930* (Bloomington, IN: Indiana University Press, 2007), 125–138.

Tanneberger, Franziska and Vera Schroeder: *Das Moor: Über eine faszinierende Welt zwischen Wasser und Land und warum sie für unser Klima so wichtig ist* (Munich: dtv, 2023).

TeBrake, William N.: *Medieval Frontier: Culture and Ecology in Rijnland* (College Station: Texas A&M University Press, 1985).

Thiel, Eric: "The Power Industry in the Soviet Union," *Economic Geography* 27, no. 2 (1951): 107–122.

Tirado, Isabel A.: "The Village Voice: Women's Views of Themselves and Their World in Russian *Chastushki* of the 1920's," *The Carl Beck Papers in Russian & East European Studies*, no. 1008 (Pittsburgh: The Center for Russian & East European Studies, 1993).

Tomoff, Kiril: "The Role of Forests in Witte's Industrialization Drive," *Russian History* 22, no. 3 (1995): 249–283.

Toomey, Lily: "'A National Resource of Primary Importance'. Transforming Ireland's Peatlands in the Twentieth Century," PhD Dissertation, Trinity College Dublin (2025).

Tsing, Anna Lowenhaupt: *The Mushroom at the End of the World: On the Possibility of Life in Capitalist Ruins* (Princeton: Princeton University Press, 2015).

"Tsirkuliar Otdela Okhrany Truda NKT SSSR o zhilishchnykh usloviiakh rabochikh na torfo-razrabotkakh," *Ėkonomicheskaia Zhizn'* (January 27, 1924).

Turetsky, Merritt R. et al.: "Global Vulnerability of Peatlands to Fire and Carbon Loss," *Nature Geoscience* 8, no. 1 (2015): 11–14.

Unger, Richard W.: "Energy Sources for the Dutch Golden Age: Peat, Wind, and Coal," *Research in Economic History: A Research Annual* 9 (1984): 221–253.

Unger, Richard W. and John Thistle: *Energy Consumption in Canada in the 19th and 20th Centuries* (Naples: Consiglio Nazionale delle Ricerche, 2013).

Vergara, Germán: *Fueling Mexico: Energy and Environment, 1850–1950* (Cambridge: Cambridge University Press, 2021).

Venovcevs, Anatolijs: "Living with Socialism: Toward an Archaeology of a Post-Soviet Industrial Town," *The Extractive Industries and Society* 8, no. 4 (2021): 100835.

Vompersky, S. E., et al.: "Estimation of Forest Cover Extent over Peatland and Paludified Shallow Peatlands in Russia," *Contemporary Problems of Ecology* 4, no. 7 (2011): 734–741.

Vozbrannaya, A., et al.: "After Wildfires and Rewetting: Results of 15+ Years' Monitoring of Vegetation and Environmental Factors in Cutover Peatland," *Diversity* 15, no. 3 (2023), doi: https://doi.org/10.3390/d15010003.

de Vries, Jan and Ad van deer Woude: *The First Modern Economy: Success, Failure and Perseverance of the Dutch Economy, 1500–1815* (Cambridge: Cambridge University Press, 1997).

Waddington, J. M., et al.: "Cutover Peatlands: A Persistent Source of Atmospheric CO_2," *Global Biogeochemical Cycles* 16, no. 1 (2001): 1–7, https://doi.org/10.1029/2001GB001398

Warde, Paul: *Energy Consumption in England and Wales, 1560–2000* (Naples: Consiglio Nazionale delle Ricerche, 2007).

"Firewood Consumption and Energy Transition: A Survey of Sources, Methods and Explanations in Europe and North America," *Historia Agraria* no. 77 (2019): 7–32.

Warde, Paul, Libby Robin, and Sverker Sörlin: *The Environment: A History of the Idea* (Baltimore: Johns Hopkins University Press, 2018).

Warde, Paul and Tom Williamson: "Fuel Supply and Agriculture in post-Medieval England," *The Agricultural History Review* 62, no. 1 (2014): 61–82.

Warner, Barry G. and Pierre Buteau: "The Early Peat Industry in Canada, 1864–1945," *Geoscience Canada* 27, no. 2 (2000): 57–66.

Weiner, Douglas R.: *Models of Nature: Ecology, Conservation and Cultural Revolution in Soviet Russia. With a New Afterward* (Pittsburgh: University of Pittsburgh Press, 2000).

"The Predatory Tribute-Taking State: A Framework for Understanding Russian Environmental History," in Edmund Burke III and Kenneth Pomeranz, eds., *The Environment and World History* (Berkeley: University of California Press, 2009), 276–315.

Wengle, Susanne A.: *Black Earth, White Bread: A Technopolitical History of Russian Agriculture and Food* (Madison, WI: University of Wisconsin Press, 2022).

Whatmore, Sarah: "Materialist Returns: Practising Cultural Geography in and for a More-Than-Human World," *Cultural Geographies* no. 13, no. 4 (2006): 600–609.

Wilkinson, S. L. et al.: "Wildfire and Degradation Accelerate Northern Peatland Carbon Release," *Nature Climate Change* 13 (2023): 456–461.

Winkler, Martina: "'Mein Besitz, Landgut, Erbland, Dorf oder wie auch immer Du es nennen möchtest': Eine russische Begriffsgeschichte des Eigentums," *Jahrbücher für Geschichte Osteuropas* 60, no. 3 (2012): 321–349.

Wilson, David: *The Demand for Energy in the Soviet Union* (London: Croom Helm, 1983).

Witze, Alexandra: "The Arctic Is Burning Like Never Before: And That's Bad News for Climate Change," *Nature* 585 (2020): 336–337.

Wöbse, Anna-Katharina: "Vom Ödland zum Reservoir des Reichtums. Die Neudeutung europäischer Sümpfe." *Ilinx—Berliner Beiträge zur Kulturwissenschaft* 5 (2020): 113–125.

World Energy Council: *World Energy Resources: 2013 Survey* (London: World Energy Council, 2013).

Wrigley, Charlotte: *Earth, Ice, Bone, Blood: Permafrost and Extinction in the Russian Arctic* (Minneapolis: University of Minnesota Press, 2023).

Wrigley, E. A.: *Energy and the English Industrial Revolution* (Cambridge: Cambridge University Press, 2010).

York, Richard and Shannon Elizabeth Bell: "Energy Transitions or Additions? Why a Transition from Fossil Fuels Requires More Than the Growth of Renewable Energy," *Energy Research & Social Science* no. 51 (2019): 40–43.

Yurchak, Aleksei: *Everything was Forever, Until It was No More: The Last Soviet Generation* (Princeton: Princeton University Press, 2006).

Zhilin, Mikhail G.: "Early Mesolithic Hunting and Fishing Activities in Central Russia: A Review of the Faunal and Artefactual Evidence from Wetland Sites," *Journal of Wetland Archaeology* 14, no. 1 (2014): 91–105.

Zhuravlev, S. V.: "AvtoVAZ nachala 1990-kh gg. Istoriia privatizatsii. Kak podelit' 'Shkuru neubitogo medvedia'?," *Sotsial'naia Istoriia. Ezhegonik* (2007): 103–148.

Index

Studies in Environment and History

Other Books in the Series *(continued)*:

Edmund Russell *Greyhound Nation: A Coevolutionary History of England, 1200–1900*
Timothy J. LeCain *The Matter of History: How Things Create the Past*
Ling Zhang *The River, the Plain, and the State: An Environmental Drama in Northern Song China, 1048–1128*
Abraham H. Gibson *Feral Animals in the American South: An Evolutionary History*
Andy Bruno *The Nature of Soviet Power: An Arctic Environmental History*
David A. Bello *Across Forest, Steppe, and Mountain: Environment, Identity, and Empire in Qing China's Borderlands*
Erik Loomis *Empire of Timber: Labor Unions and the Pacific Northwest Forests*
Peter Thorsheim *Waste into Weapons: Recycling in Britain during the Second World War*
Kieko Matteson *Forests in Revolutionary France: Conservation, Community, and Conflict, 1669–1848*
Micah S. Muscolino *The Ecology of War in China: Henan Province, the Yellow River, and Beyond, 1938–1950*
George Colpitts *Pemmican Empire: Food, Trade, and the Last Bison Hunts in the North American Plains, 1780–1882*
John L. Brooke *Climate Change and the Course of Global History: A Rough Journey*
Paul Josephson *An Environmental History of Russia*
Emmanuel Kreike *Environmental Infrastructure in African History: Examining the Myth of Natural Resource Management*
Gregory T. Cushman *Guano and the Opening of the Pacific World: A Global Ecological History*
Sam White *The Climate of Rebellion in the Early Modern Ottoman Empire*
Edmund Russell *Evolutionary History: Uniting History and Biology to Understand Life on Earth*
Alan Mikhail *Nature and Empire in Ottoman Egypt: An Environmental History*
Richard W. Judd *The Untilled Garden: Natural History and the Spirit of Conservation in America, 1740–1840*
James L. A. Webb, Jr. *Humanity's Burden: A Global History of Malaria*
Myrna I. Santiago *The Ecology of Oil: Environment, Labor, and the Mexican Revolution, 1900–1938*
Frank Uekoetter *The Green and the Brown: A History of Conservation in Nazi Germany*
Matthew D. Evenden *Fish versus Power: An Environmental History of the Fraser River*
Alfred W. Crosby *Ecological Imperialism: The Biological Expansion of Europe, 900–1900, second edition*
Nancy J. Jacobs *Environment, Power, and Injustice: A South African History*
Edmund Russell *War and Nature: Fighting Humans and Insects with Chemicals from World War I to Silent Spring*
Adam Rome *The Bulldozer in the Countryside: Suburban Sprawl and the Rise of American Environmentalism*

Judith Shapiro *Mao's War against Nature: Politics and the Environment in Revolutionary China*
Andrew Isenberg *The Destruction of the Bison: An Environmental History*
Thomas Dunlap *Nature and the English Diaspora*
Robert B. Marks *Tigers, Rice, Silk, and Silt: Environment and Economy in Late Imperial South China*
Mark Elvin and Tsui'jung Liu *Sediments of Time: Environment and Society in Chinese History*
Richard H. Grove *Green Imperialism: Colonial Expansion, Tropical Island Edens and the Origins of Environmentalism, 1600–1860*
Thorkild Kjærgaard *The Danish Revolution, 1500–1800: An Ecohistorical Interpretation*
Donald Worster *Nature's Economy: A History of Ecological Ideas, second edition*
Elinor G. K. Melville *A Plague of Sheep: Environmental Consequences of the Conquest of Mexico*
J. R. McNeill *The Mountains of the Mediterranean World: An Environmental History*
Theodore Steinberg *Nature Incorporated: Industrialization and the Waters of New England*
Timothy Silver *A New Face on the Countryside: Indians, Colonists, and Slaves in the South Atlantic Forests, 1500–1800*
Michael Williams *Americans and Their Forests: A Historical Geography*
Donald Worster *The Ends of the Earth: Perspectives on Modern Environmental History*
Robert Harms *Games against Nature: An Eco-Cultural History of the Nunu of Equatorial Africa*
Warren Dean *Brazil and the Struggle for Rubber: A Study in Environmental History*
Samuel P. Hays *Beauty, Health, and Permanence: Environmental Politics in the United States, 1955–1985*
Arthur F. McEvoy *The Fisherman's Problem: Ecology and Law in the California Fisheries, 1850–1980*
Kenneth F. Kiple *The Caribbean Slave: A Biological History*

For EU product safety concerns, contact us at Calle de José Abascal, 56–1°, 28003 Madrid, Spain or eugpsr@cambridge.org.

www.ingramcontent.com/pod-product-compliance
Lightning Source LLC
Chambersburg PA
CBHW051256200926
51612CB00002B/60
9781009603089